How to Build PATIO ROOFS

By the editors of Sunset Books
and Sunset Magazine

LANE PUBLISHING CO. · MENLO PARK, CALIFORNIA

Edited by John Gillespie

Design: John Flack, JoAnn Masaoka

Illustrations: Ted Martine

Cover photograph: George Selland, Moss Photography

Editor, Sunset Books: David E. Clark

Eighth printing May 1982

 Library of Congress No. 73-89579.
ISBN 0-376-01455-5. Lithographed in the United States.

Contents

Between you and the sun	5
Building from the ground up	15
Patterns in reed and bamboo	27
Screens of fabric and metal	33
Lath, batten and lumber	39
The canvas cover	47
Glass, plastic and aluminum	53
Louvers for sun or shade	65
Eggcrate, the versatile roof	71
Solid roofs	77
Index	80

Special Features

Defending against sun and wind	9
Heavy snow loads: special problems	11
Patio roof materials chart	12
How to construct a patio overhead	25
How to make box beams	29
Ideas for reed and bamboo	31
Screens make you unbuggable	37
Five ways to design a roof	44
Three ways to canvas	51
The ventilated patio	62
Controlling the sun and shade	68
Vines for patio roofs	72
Versatile eggcrate	74
A convertible lanai	79

INTRODUCTION:

Between you and the sun

The idea of building a patio roof promises greater versatility for your outdoor living room — a sheltered play area for children, a place for family fun, barbecues, entertaining guests, or relaxing.

Anyone who has roofed over his patio can tell you what a transformation this simple construction job can make. A desertlike expanse of hot paving becomes a cool inviting shelter. A damp and windy tundra changes to a light, warm haven for daytime or evening recreation. Rooms in the house that had been uninhabitable during the summer become cool and pleasant under a new roof.

Unfortunately, it is also possible to build a patio roof that will increase the problem you are attempting to solve or substitute a new headache for the one you have eliminated. With the wrong type of roofing or with inadequate planning, what was merely an uncomfortably warm patio may be changed into an unbearable kiln. A walled-in patio may become stuffier than ever when you put a roof over it. Or the desperately needed shade is cast over an area where it is not needed or during a time of day when no one uses the patio. Without careful planning, a rainy patio may be covered over with a roof that drips from condensed moisture.

Obviously, your outdoor center will need forethought. You must decide whether your roof will serve you better built against the house or a short distance away from it, near the swimming pool or on the back of the lot. Convenience to cooking facilities is important to location. Unless your backyard chef plans to prepare meals on the patio itself or nearby, the patio ought to be near the kitchen. If your outdoor area is not conveniently located, too many shuttle trips to the house for supplies can turn an afternoon or evening's fun into tedious drudgery.

Planning also involves the structure itself—detailed physical plans, tools, materials, as well as the knowledge of how to build the supporting framework. An attractive, sturdy structure doesn't just happen — it's planned.

SUNLIGHT IS FILTERED through roof of finished lath that is supported by fence and bamboo post.

LET CLIMATE BE YOUR GUIDE

A withering sun can send you in search of a cool retreat as quickly as snow or rain can drive you indoors. Sweltering or inclement weather does not invite the idea of a patio party. If you have lived in your present climate for a number of seasons, you are already familiar with its benefits and hazards, but if you have recently moved to a region that is new to you, you may want to know what to expect in the coming seasons.

Because day-to-day weather conditions are highly localized, forecasts based on rules-of-thumb, the neighborhood "expert," or theories without scientific basis for support are generally unreliable. Concise and comprehensive climate information is readily available.

Send fifteen cents to: NOAA (National Oceanic and Atmospheric Administration), National Climate Center, Federal Building, Asheville, North Carolina 28801. Ask for the current Local Climatological Data, annual issue, naming your area. If you want monthly information, the subscription cost is $2.00 per year and includes the annual issue.

Accurate weather information is also available from the following sources:

- U. S. Weather Bureau offices
- Federal Aviation Administration branches and local airport operation offices.
- National Park Service, U. S. Forest Service, and state forestry district offices and ranger stations
- U. S. Coast Guard stations and district offices
- Public power and utility companies
- Municipal water districts
- City, county, and state road and highway maintenance departments (snow information in areas with heavy winters)
- College and university meteorology departments
- County farm bureaus and agricultural extension agents.

Your relation to the sun

Theoretically, a patio facing north is cold because it never receives the sun. A south-facing patio is warm because from sunrise to sunset, the sun never leaves it. On the east side a patio is cool, receiving only the morning sun. The west side story is withering hot!

In the United States (excluding Hawaii) and Canada, the sun rises and sets in a progressive pattern farther north each day between December 21 and June 21 of the following year; between June 21 and December 21 the pattern reverses as the sun edges progressively southward.

Three points determine the daily path of the sun: sunrise, the noon position and sunset. Knowing the location of these points, you can plot the sun's path above your property at all seasons. This information is important in patio planning because the sun's path combined with wind direction govern the warmth or coolness of your outdoor room.

To find your patio's relation to the sun, locate your house on the map shown below. Next note the directions of sunrises and sunsets in relation to your home (shown in the chart for your area). To make on-the-spot observations, go to the site you have selected for your patio. Stand where you expect to spend most of your time. To predict whether the spot you're standing on will be in sun or shade during any given season, raise your eyes to the sun's noon position above the horizon shown on a chart for that season. If you see open sky, that spot would be sunny during that season; if you see anything blocking open sky (trees, buildings, hills, or other obstructions), that spot would be shaded.

Generally, your patio's temperature will follow the north, south, east, west rule mentioned above. Because it is a general rule, however, it is best to get specific climate information from one of the sources listed on this page.

Exceptions to this north, south, east, west idea are climates where extreme summer and winter temperatures are a measurable certainty. For example, in Phoenix, Arizona, with mid-July temperatures ranging upwards from 100 degrees, a north patio cannot be described as cold. Nor in San Francisco, can a south patio be considered warm, or a west patio hot, when a stiff ocean breeze has escorted the chilling fog to your patio party.

Plan your patio location carefully. First study the area chart below, the sun chart on page 7, and the wind information on page 8. You can then choose a patio location that is enjoyable over a longer outdoor season. Maximum benefit from your patio roof may also depend upon your arrangement of the area immediately surrounding the patio.

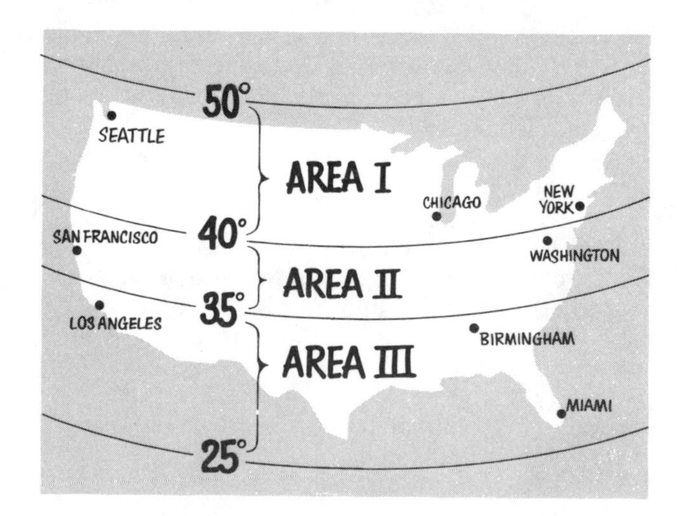

THE SUN'S PATH OVER YOUR HOUSE

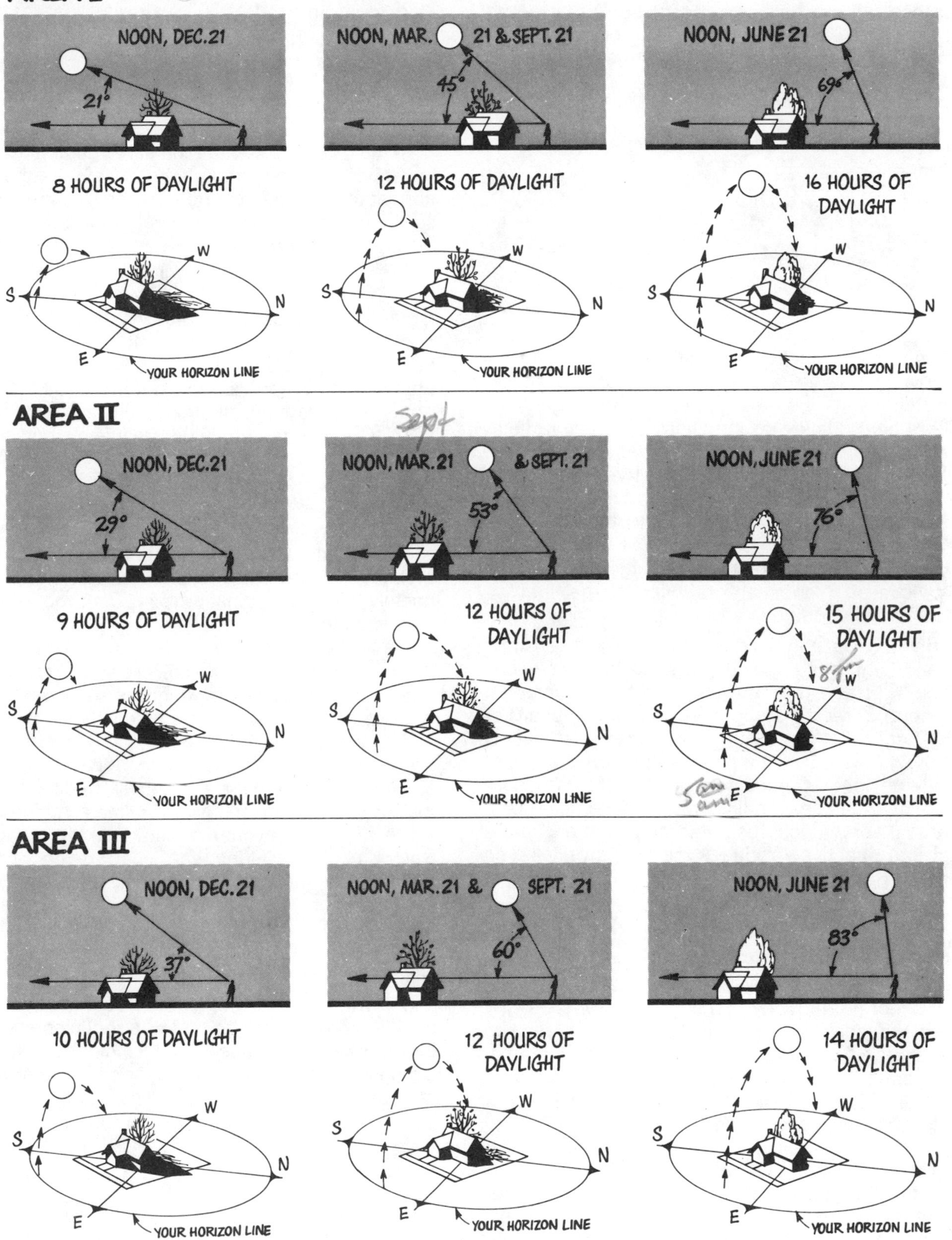

THE SUN'S ARC over your property depends upon the latitude and the season. Combining these factors, you can calculate where shade will fall across your property. Use the map on the facing page to find your area first.

Vertical screens. Your only defense against the western sun is vertical screening. Unless your climate provides cooling afternoon and evening breezes, you can expect your outdoor living area to become ovenlike, therefore uninhabitable.

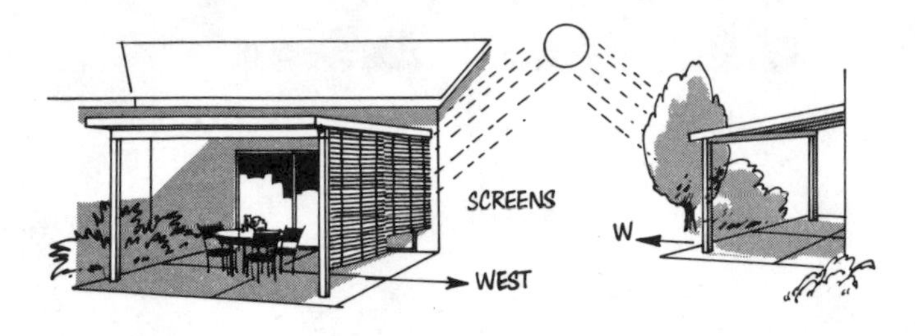

The best protection against the withering heat is a row of closely spaced tall shrubs or a lattice-and-climbing-vine arrangement. A growing screen filters heat quickly on the sunny side and radiates coolness on the shady side. Unless you buy the plants full grown (relatively expensive to do), install a temporary screen while the shrubs are growing. (See photo at right.) Alternates to the tall shrubs are a trellis and vine cover or a tall hedge. If your patio situation is complex, you may require the services of a landscape architect to get the outdoor living conditions you want.

If you choose to build a screen several possibilities in hundreds of designs are available. You can hang a screen from the patio roof, using canvas, bamboo or reed matting, plywood, plasterboard or any heat reflecting material. Another possibility is to build a wall or fence against the sun. Also, specially designed frames can be built and their contents changed seasonally, or for special occasions. Vertical screens can be made of almost any material that effectively reduces the heat's penetration of your patio.

Wind, rain, and snow

The inclement conditions of wind, rain, and snow will affect not only the location (and consequently the usefulness) of your patio but also will tend to dictate the type of roof structure you should build. The direction from which prevailing winds and storms approach is a key climate control factor in patio design, planning and location.

Wind. Of all weather's effects, wind (however light) has the largest influence on day-to-day outdoor comfort. Wind, along with sun and shade, determines how warm or cold you feel outdoors. U. S. Army studies indicate that if you are sunbathing in a still-air temperature of 70° and a 5 mile-per-hour breeze approaches, you will experience a chill temperature equal to 50° without wind. In a 20 mile-per-hour wind at 70°, you will feel a drop in temperature of about 30°.

Prevailing winds should be considered in planning both the design and location of your patio. Extra strength may be necessary for a patio roof facing high winds; otherwise the formidable lift generated may raise it off the substructure. Knowing wind directions and strength may also be important for you in deciding whether you need wind screens, since they usually require structural attachment to the framing.

Three different types of winds affect patio locations in some regions: annually prevailing winds, highly localized seasonal breezes (daily, late afternoon, summer, near beaches), and occasional high-velocity winds (hurricanes off the Atlantic seaboard, mid-western thunderstorm squalls, "Santa Ana's" of Southern California). Official sources can usually tell you the direction, frequency, and average velocity of these winds and breezes as they may apply to your locality.

Rain. Unless accompanied by high winds, seasonal rainfall does not generally create serious problems. If you apply exterior preservatives adequately to guard against warping, your roof will not be damaged.

Snow. In areas of snow fall, roofs are most frequently damaged by the weight of snow gathering in drifts or avalanching off gabled roofs. Snow is deceptively heavy. When piled 5 to 6 feet on a flat overhead, snow weighs 80 to 100 pounds per square foot, twice the maximum allowed by municipal ordinances in many mild-climate regions. In areas of extremely heavy snowfall, the standing weight of snow alone can snap rafters, crush beams, and cause collapse of the substructure. Improperly designed construction has little chance to survive. Even in areas of relatively light snowfalls, your roof should be carefully designed to withstand snowloads.

Snow tends to drift against vertical obstructions tumbling down from the part of a roof slope facing the approach direction of winter storms. Plan to strengthen those parts of the supporting framework of your roof.

Is your patio pocketed by the house?

An L or U patio area formed by the walls of the house creates a roofing situation requiring specialized knowledge. Because an incorrectly installed patio roof is bound to cause you problems,

DOUBLE PROTECTION shields patio on breezy days. Doors slide shut; canvas draws across entry. Landscape Architect: William Kapranos.

WOOD MOSAIC SCREEN. Pieces are fastened to ½-inch plywood backing.

Defending against sun and wind

ABOVE: bamboo fence provides privacy; aluminum screen protects plants; redwood roof shades deck. LEFT: the 6 by 6 posts support pattern of 2 by 4s and 1 by 2s mounted on framework of 4 by 6s.

a consultation with an architect or engineer is recommended. An L with west and north-facing walls will need secure sun protection for the west-facing wall, but if the north-facing wall has a window, you must consider the effect a roof will have on the windowed room, particularly in winter.

As the drawing illustrates, you can build an adjustable or movable roof so that light can be admitted when needed. You can use lightweight materials—panels, reed, canvas, bamboo—that can be stored during winter. Another treatment is to plant a deciduous vine cover (see page 72) or simply to leave an open space between the roof covering and the north-facing wall:

If the patio is in a U or a fully enclosed court, your biggest task will be to maintain adequate ventilation. Should you fail to do this, your new roof will cover a hot box. Ventilation is essential; be certain that you do not seal in the area:

Closely spaced lath, shrimp net, reed, bamboo, or other similar materials should serve you best. Heat-transmitting coverings such as glass, fiberglass, or plastic should be installed with caution or avoided entirely. If installed, they should be provided with lath, reed, bamboo, louvers, or another sun-filtering cover, fitted into place above the glass. You would be wise to consult a professional before installing heat-transmitting materials.

YOUR PATIO'S FUTURE

Patios have a way of evolving, almost imperceptibly, into enclosed rooms, especially after the roof is constructed. When you build a roof over your patio, you are adding a permanent improvement to your home. Therefore, before going down to the lumber yard "to check on a few prices," it would be wise to give some thought to your patio's future development.

Will you want to enclose it entirely? If you are planning for a light, temporary covering, do you foresee eventual conversion to a solid, more substantial roof? What roughed-in preparation for future development can you make when you install the roof? Will the roof's supporting framework serve as a bearing wall later, or will you have to rip the whole thing out and start from scratch? The time to figure out these problems is before you put up a single 4 by 4. Not all of these improvements will affect your construction of the roof—but some will, and you can avoid some serious problems later by thinking the whole project through now.

Arranging for professional help

Having completed preliminary planning and site location, turn to the next decision: how much of the job should you undertake? The decision usually depends on your budget, time available, carpentry skills, and the extent of the work involved.

Building a basic patio roof on a conventional substructure should offer no serious problems to the home craftsman who has a few free weekends or evenings. Certain terrain conditions and some types of roofs, though, will require partial or complete professional handling.

For architectural aid, choose a residential architect who practices locally. Chances are good that he will be familiar with local materials, contractors, and conditions of soil, climate, and so forth. It's wise to take several bids and ask for references from former clients. The fee, based on a percentage of construction costs, will depend upon the extent of his work. A complete service will prepare working drawings, take competing bids from contractors, assign contract, and supervise construction.

Budget may be a key factor in deciding on the degree of professional help you need in building a patio roof. A harsh reality: having a roof professionally designed and completely constructed by a contractor will probably cost you five times or more than if you do all the work yourself. One common practice—a compromise of cost extremes—is to hire a retired or independent carpenter to work with you. An added advantage of doing this is that he may provide his own tools, reducing or eliminating the cost of tool purchases or rentals.

If you decide to hire a contractor to do part or all of the work, be sure to ask several contractors to bid on the job.

MOUNTAIN CABINS AND PATIO ROOFS HAVE THE SAME SNOW PROBLEMS

Architects familiar with snow country feel that the answer for any roof shape may be a return to the metal roof. Used extensively in early cabin construction, metal roofs are now banned in most resort developments because of "unsightliness." But new color-coated roofs now being developed are attractive and highly functional. Leaks are seldom a problem with metal roofs, as the entire snow load has nothing to adhere to and usually slips off naturally.

A mountain roof contends with some unusual stresses. It must carry heavy snow loads, withstand repeated thawing and freezing, and insulate against extreme cold. Five designs used in snow country are described here. Each has its advantages and disadvantages.

The hip roof or shed roof is common in snow country and can often do well in deep snows. But there are a number of problems inherent in any roof with eaves.

Many owners of hip and shed roofs are forced to tack heat conductive wires ("heat tapes") on the eave areas to melt ice and prevent ice dams. This creates costly electricity bills, fire hazards, and a maintenance problem with the tapes, which may need replacement often.

The A-frame which is dramatic and inexpensive to build, is one answer to the problem of snow loads. But even though the A-frame's steep sides will shed snow off its upper reaches, the snow has nowhere to fall and begins piling up against the structure itself. This can create unequal lateral forces. Tons of snow from the roof, combined with snow already drifted against one side of the posts, can cave in a roof and may even push the structure off its foundation. One answer is to build the A-frame 10 or more feet off the ground.

A flat roof has a number of advantages. Some contractors recommend flat roofs for snow country over all other types, comparing the snow load on a flat roof to a person sitting in an antique chair — "If you sit up straight and don't move, the chair won't break, but if you tilt back, the legs will snap." A flat roof has only one force to contend with, and that's straight down. This requires large timbers to support the tremendous load, but saves on lineal footage of lumber.

The "no-eave" roof is often used by architects to prevent insidious leaks. Outside walls are canted to create an overhang.

Icicles often reaching as much as 20 feet to the ground are the proof that snow isn't freezing on the roof, but is running off its edge. Another advantage is that snow cascading from the roof is dumped somewhat away from the structure and doesn't pile up against it.

Slanting walls also create interesting interior spaces. But the overall design results in a somewhat more expensive design because of extra lumber, bracing, and workmanship.

A cold roof is really two roofs with a ventilated air space sandwiched between them. This air space allows the outer roof to remain at the same temperature as its snow load. Snow continues to build up naturally and stay in place all winter (without melting near the roof surface). This stops ice damming and leakage, and may prevent the sudden breaking loose of tons of snow that can sweep away or crush everything in its path. But cold roofs are expensive, essentially doubling your initial roof costs in extra labor, special design considerations, and materials.

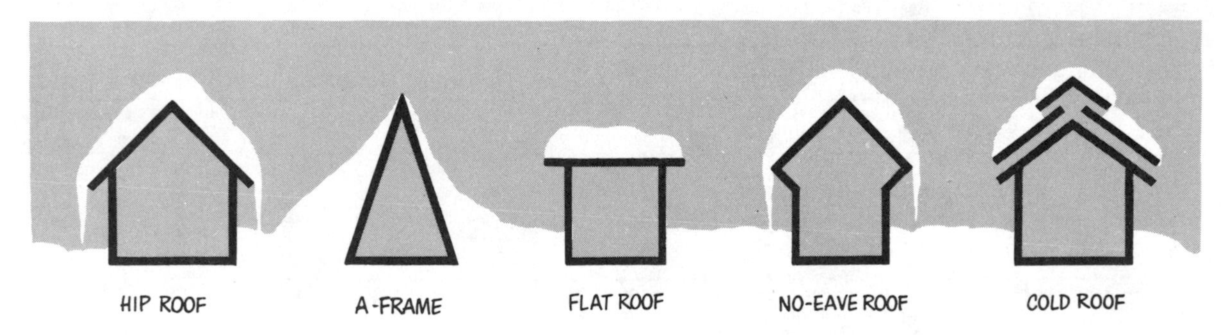

THE SAME ROOFING PRINCIPLES that would apply to a mountain cabin should be remembered when planning the basic style of a patio overhead.

COMPARATIVE SUMMARY OF PATIO ROOFING MATERIALS

Name of Material	Description	Type of Shade	Ventilation	Rain-shield	Standard Unit Sizes
Lath and Lumber					
Lath	Thin slats, rough; long life	Striped, variable	Good	None	⅜″ x 1½″ by 4, 6, 8-ft. lengths
Batten	Thin slats, rough or surfaced; long life	Striped, variable	Good	None	¼″ to ¾″ thickness by 2″ to 3″ wide to 20-ft. lengths
Woven wood panels	Flat strips woven in basket weave; long life	Dense, spotted	Fair	None	½″ by 4″ and ½″ by 6″ to 20-ft. lengths
Beanpoles	Slim, square poles; long life	Striped, variable	Good	None	1″ x 1″ by 6′ to 8′ in length
Grapestakes	Rough, split stakes; long life	Striped, variable	Good	None	2″ x 2″ and 1″ x 2″ usually 6′ long
Wood and wire utility fencing	Spaced pickets woven together with wire; long lifespan	Striped; slats widely spaced	Good	None	1′ to 4′ wide in 50-100′ rolls
Wood slats, cord woven	Thin slats tightly woven together with cord; several years life	Dense, striped	Some	None	2′ to 12′ width to any length
1 x 2s, 1 x 3s	Rough or finished lumber; long life	Striped, variable	Good	None	4′ to 20′ length
1 x 4s, 2 x 2s, 1 x 6s	Rough or finished lumber; long life	Striped, variable	Good	None	4′ to 20′ length
Eggcrate	Grid formed of dimension lumber; long life	Slight, early and late in day	Good	None	To fit
Aluminum lath	Light, prefinished, used like lath; long life	Striped, variable	Good	None	2″ to 8″ by any length
Blinds, Screens, Meshes					
Bamboo blinds	Strips woven together with cotton cord; life of 2 to 4 years	Good, soft broken	Good	None	2′ to 10′ wide by 6′ to 12′ in length
Reed Screen	Reeds woven together with stainless steel wire; life of 2 to 5 years	Good, soft broken	Good	None	6′4″ wide by 25′ long
Flyscreen					
Aluminum	Long life; some corrosion near seashore; bulges if struck	Slight	Good	None	16″ to 72″ wide in rolls of varying length
Bronze	Bright color; water running from it can cause stain	Slight	Good	None	16″ to 72″ wide in rolls of varying length
Galvanized steel	Standard window screen; rusts without periodic painting	Slight	Good	None	16″ to 72″ wide in rolls of varying length
Plastic	Unaffected by salt air; difficult to install. Shrinkage big problem	Slight	Good	None	24″ to 48″ wide in rolls
Glass fiber	Vinyl-coated; strong; won't dent or corrode	Slight	Good	None	22″ to 84″ wide in rolls
Saran shade cloth	Plastic mesh; many densities; use in mild climates only	30% to solid shade, depending on density of weave	Good to fair	None	24″ to 60″ widths; some weaves up to 20′ wide
Plastic-encased screen	Light, flexible; wire embedded in plastic; short life	None, but sun is diffused	Heat trap unless vented	None	36″ wide in rolls
Louvered screen					
Aluminum	Sheet aluminum with tiny louvers cut into it. A special order item.	Good if louvers are oriented properly	Good	None	18″ to 48″ wide in rolls
Aluminum-and-plastic	Plastic-coated aluminum wires; neutral gray	Good if louvers are oriented properly	Good	None	24″ to 48″ wide in rolls

Name of Material	Description	Type of Shade	Ventilation	Rain-shield	Standard Unit Sizes
Glass					
Glazed hotbed sash	Panels of glass mounted in special sash, unlimited life if wood kept painted	None	None, if not vented	Good	3′ x 3′ to 6′ panels
Window glass	Used in greenhouse type construction. Unlimited life if not shattered.	None	None, if not vented	Good	16″ x 18″ or 16″ x 20″ commonly used
Wire glass	Strong, thick, required for overheads or skylights in many areas. Lasts indefinitely.	None	None, if not vented	Good	Up to 4′ x 12′
Solid Panels					
Exterior plywood	Won't de-laminate, but may check if not treated occasionally	Total	None, if not vented	Good	4′ x 8′
Tempered hardboard	Same	Total	Same	Good	4′ x 8′
Asbestos cement	Indestructible; plain or color; must be drilled before nailing	Total	Same	Good	Flat—4′ x 8′, 10′, 12′ Corrugated—42″ wide x 8′, 10′, 12′
Aluminum	Flat, ribbed, corrugated, V-crimp; unlimited life; available plain or in colors	Total	None, if not vented	Good	26″ to 49¾″ wide, 6′, 8′, 10′, 12′ long
Steel, galvanized	Corrugated; rusts quickly near seashore	Total	Same	Good	27½″ by 6′, 8′, 10′, 12′ in length
Translucent plastic	Many colors, degrees of translucency; flat, corrugated; fiberglass-reinforced types excellent; vinyl good (also available opaque); plain polyester doesn't last long	Light to heavy	Heat trap unless vented	Good	Panels: 26″ to 51½″ wide, 8′ to 20′ long. Rolls: to 50′ long; corrugated, 40″ wide; flat, 36″ wide
Felt-coating-gravel (built-up roofing—3-ply)	Layers as follows: roof coating on roof sheathing; felt; roof coating; felt at right angles to first layer; roof coating; felt running same direction as first layer; roof coating; gravel—100 lbs. per 100 sq. ft.	Total	Same	Good	15# felt 3′ wide: 324 sq. ft. rolls. Roof coating: 1 or 5-gallon cans. Gravel by yard or ton
Fabrics					
Cotton canvas	Many kinds; painted, dyed, vinyl-coated, unfinished; 1 to 8 years life. (Dacron canvas lasts longer, but quite expensive.)	Light to dense; dep. on finish	None	Good	Standard width 31″; other widths available
Canvas strips for woven overhead	Cut from 31-inch roll, strips usually 7″ to 9″ wide	Varies with weave; squares of light	Good	None	Custom made
Burlap	Rough, dark fabric; 2 to 4 years life	Medium to heavy	Fair	None	Bolts 36″-38″ wide
Cotton sheeting	Light, undyed fabric	Light, direct rays blocked	Poor	None	Bolts 2′-5′ wide
Perforated Panels					
Aluminum patterned, corrugated or flat	Embossed metal; unlimited life	Pinpoints of light from small holes 1″ apart	Good	None	3′ x 3′, 4′ x 8′ panels
Tempered hardboard	Pressed wood panels; long life	Same	Good	None	4′ x 4′—4′ x 8′ 2′ widths, 3′, 4′, 6′ lengths
Expanded metal Steel	Galvanized and black (diamond lath) types. Metal punched, stretched to form grid; lasts longer if painted	Some, spotted	Good	None	Galv.—3′ x 8′—4′ x 8′ Black—27″ x 8′
Aluminum	Same, but no paint needed	Same	Same	None	Panels: 26″ to 51½″

Building from the ground up

The basic components and a step-by-step method

EGGCRATE, IMAGINATION create Oriental garden scene.

Building a roof for your patio is a simple job once you decide on the important details. If you are handy with a hammer and saw and choose to do all or any portion of the work yourself, you must obtain a building inspector's approval of your plans. Before starting to design your shelter, it is wise to check on setback and height restrictions. Some communities prohibit building any overhead structure closer than 5 feet to the property line. If you build your shelter right to the fence, the building inspector may force you to dismantle it. Then, too, a building code violation might cloud the title to your property at some future time if you decide to sell the property.

To apply for a building permit, a simple plan sketch showing relationship to property lines, lumber dimensions, framing details, and piers or foundations is normally all that is required.

Design your structure to satisfy minimum requirements so that:

1. It will meet requirements for approval.

2. It can support a heavy weight—a roofer, the TV repairman, or snow (depending on your climate). You may have to perch on the structure while you are building it or even climb up there later for repair work.

3. It will withstand wind pressure. Even if you do not live in a windy area, you should design your overhead so it can withstand the lifting force of an extraordinary seasonal wind.

4. It should not have to be rebuilt or strengthened if you (or a future owner) decide to close in the area to create a lanai or an extra room.

In climates where the weight of snow must be considered, beams and rafters should be carefully calculated to allow for this extra load. If you plan to build an overhead for a mountain cabin in a snow area, you will need expert advice because unless it doubles as a ski cabin, your structure will be unattended during the winter, and snow may pack heavily on the overhead. Even an open

GABLES of shingled plywood provide ventilation for this patio area.

frame may be subjected to crippling pressure from packed snow.

THE BASIC PARTS

Footings, posts, beams, rafters, and a covering (optional) are the usual components of a patio overhead. Most structures transfer the roof weight to the rafters, then to the beam, then down the post to the pier and/or footing. We treat the basic parts in the order that you will build your structure, from the ground up.

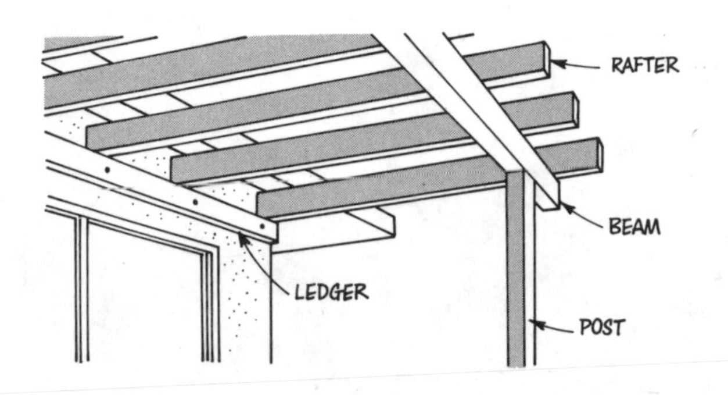

Setting posts for support

For most garden structures, 4 by 4 heartwood posts are sufficiently large. For large structures, bigger posts are sometimes necessary to support heavier roof loads and still maintain wide post spacing.

Posts in the ground. The simplest method of anchoring a supporting post is to set it directly in the ground and firmly tamp the earth around the post until it has sufficient rigidity (see Footings). If soil is sandy or unstable, you should pour a concrete collar around the post on top of the tamped earth.

The depth to which posts should be set depends on the soil conditions and wind load. Adequate for most 8 to 10-foot-high structures is 36 inches.

Use foundation grade redwood, cedar, or a pressure preservative treated wood for all ground-anchored support posts and for any other wood construction that comes within 6 inches of the ground.

When supporting posts are set in the ground, the floor of the structure may be a deck, gravel, paving blocks, grass, or earth. Ordinarily, if a concrete slab is to be poured for the floor, the posts are anchored to the slab rather than set in the ground.

Posts on concrete. Three common methods are illustrated. Patented post anchors of many types, available at most local building supply dealers, may be imbedded in the concrete (see illustrations on opposite page). This method provides positive anchorage of the posts to the concrete. The nailing block method allows the post to be toenailed to a redwood block that has been set in concrete.

When a concealed anchorage is desired, the drift pin is often used. A moisture barrier—a piece of flat metal or heavy asphalt paper—should be left between the bottom of the post and the concrete surface to avoid the accumulation of moisture and dirt.

If the structure is to be built above an existing deck, the posts may be placed over the existing support members of the deck. The posts should be firmly anchored with angle fasteners shown in the hardware drawings below.

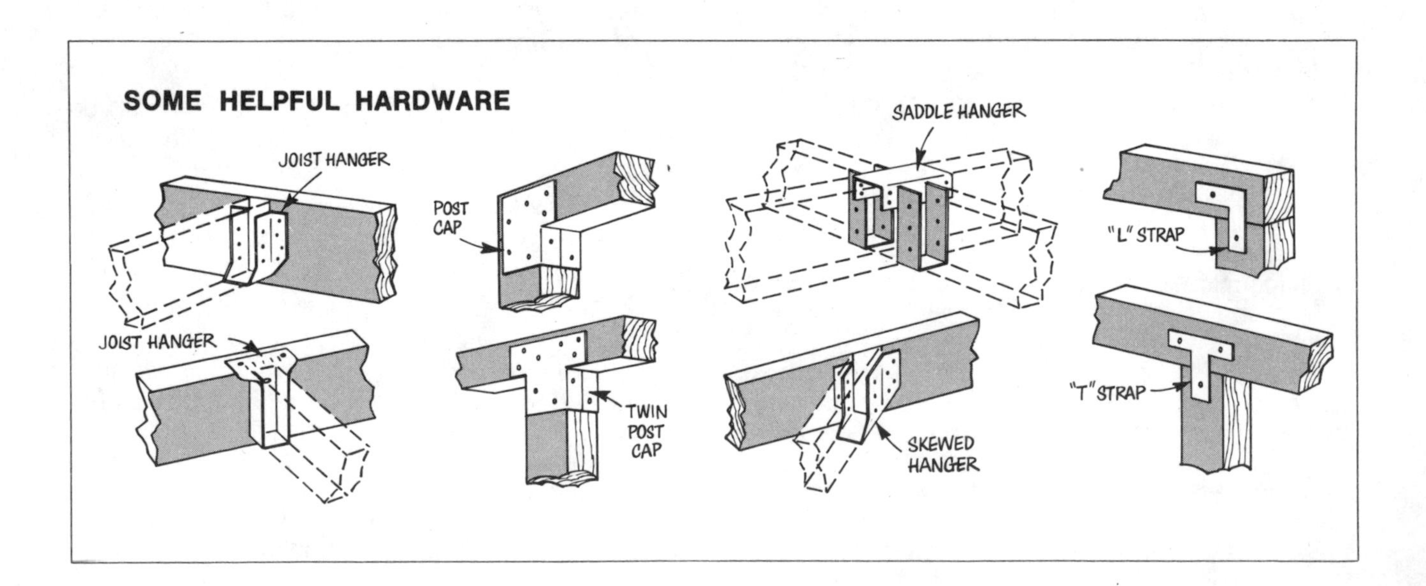

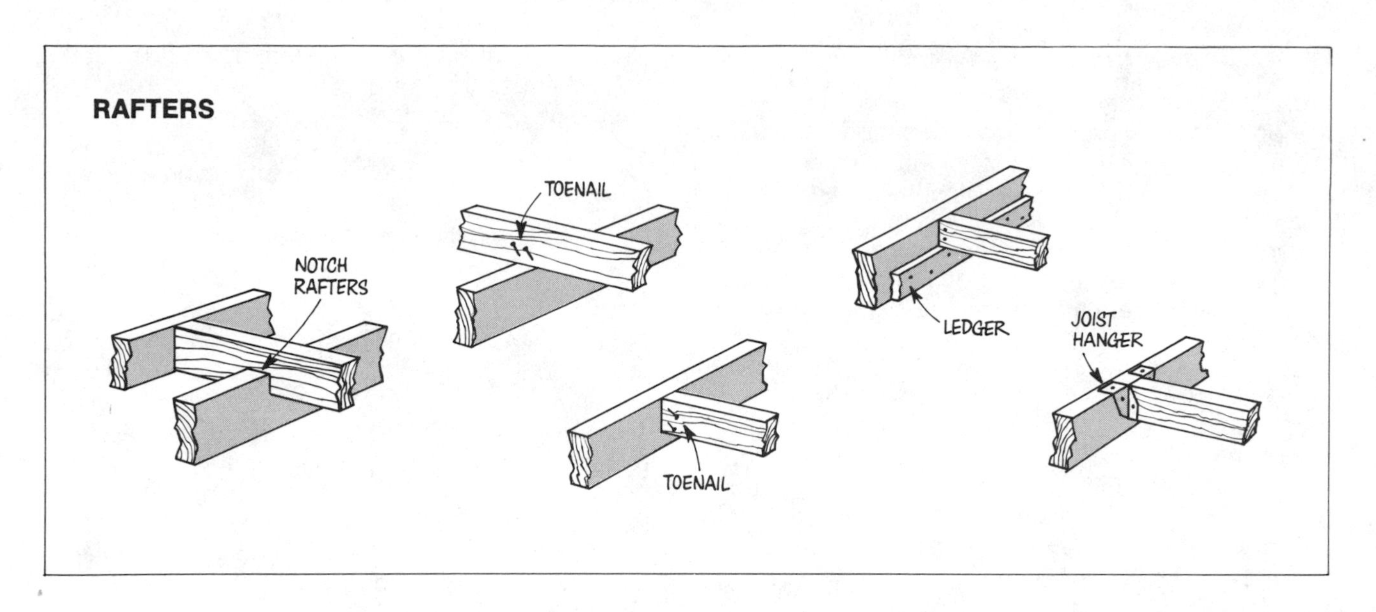
RAFTERS
NOTCH
RAFTERS
TOENAIL
TOENAIL
LEDGER
JOIST
HANGER

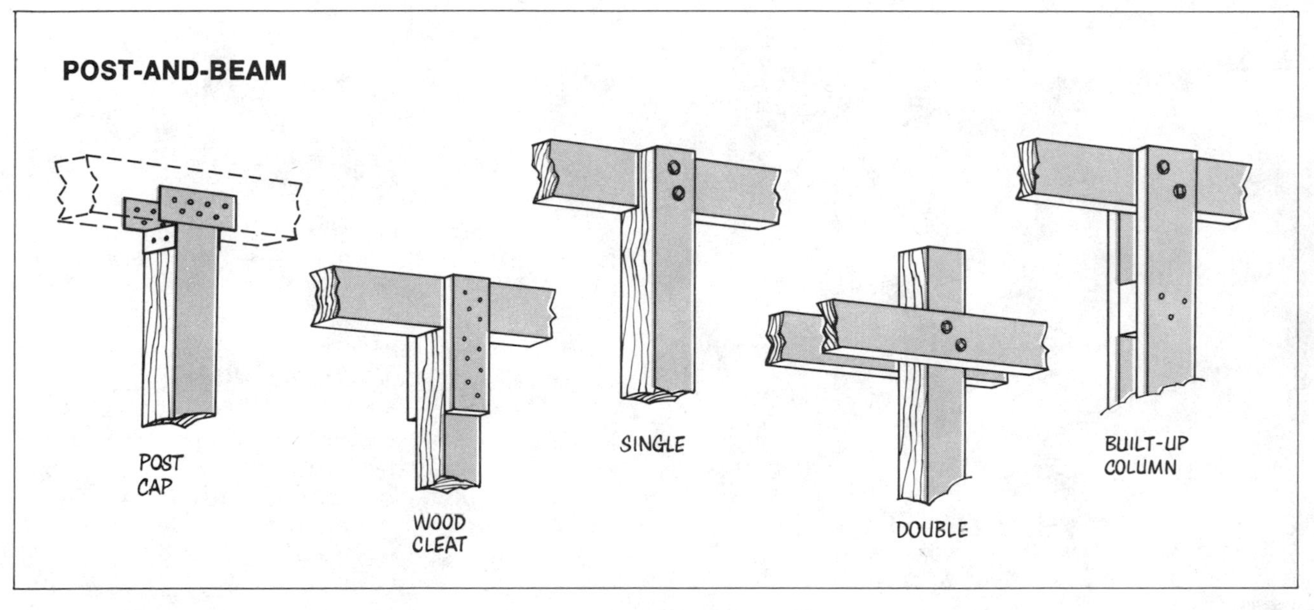
POST-AND-BEAM
POST
CAP
WOOD
CLEAT
SINGLE
DOUBLE
BUILT-UP
COLUMN

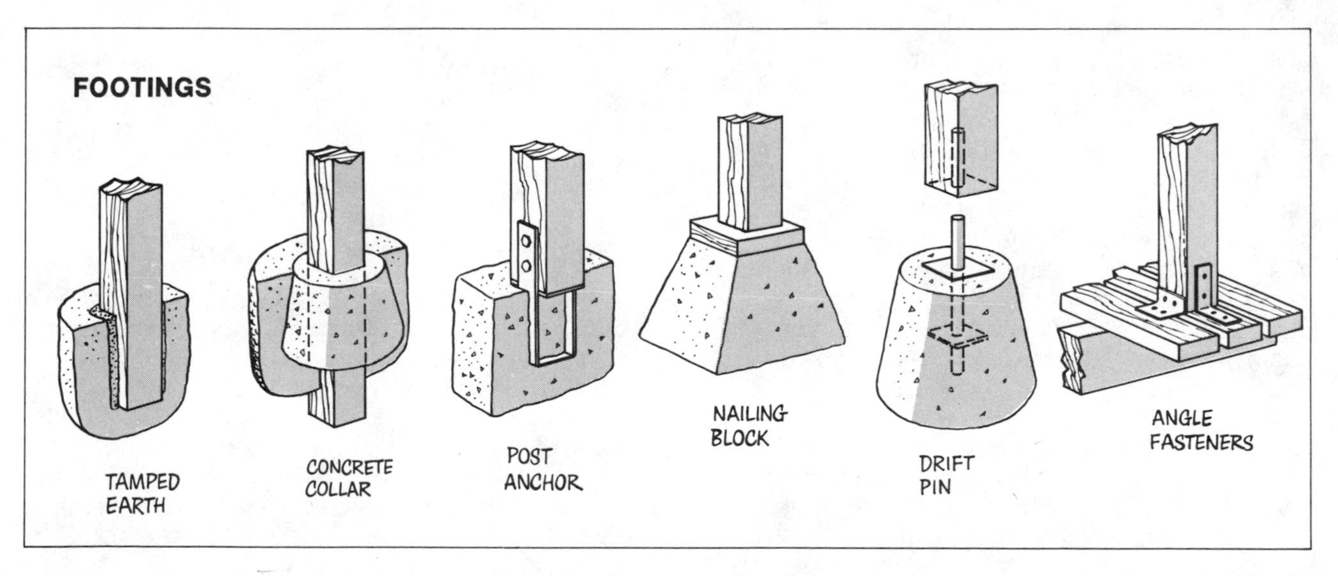
FOOTINGS
TAMPED
EARTH
CONCRETE
COLLAR
POST
ANCHOR
NAILING
BLOCK
DRIFT
PIN
ANGLE
FASTENERS

Posts and beams

From the beam the load is transferred directly to the post, then carried to the foundation or pier.

An almost universal post is the common redwood or Douglas fir 4 by 4. It will support such a heavy roof load that only in unusual situations will you need a post of larger dimensions.

If you want to be certain that your post will support its load, calculate the area of roof supported by each separate post. This area is bounded by lines drawn halfway between the post and any adjoining post or wall. Multiply the area by the roof loading figure recommended for your locality. (In moderate climates, 30 pounds per square foot will provide a safe loading figure.) A 4 by 4-inch post, 8 feet or less in length, will support better than 8,000 pounds. A 4 by 6-inch post will support better than 14,000 pounds.

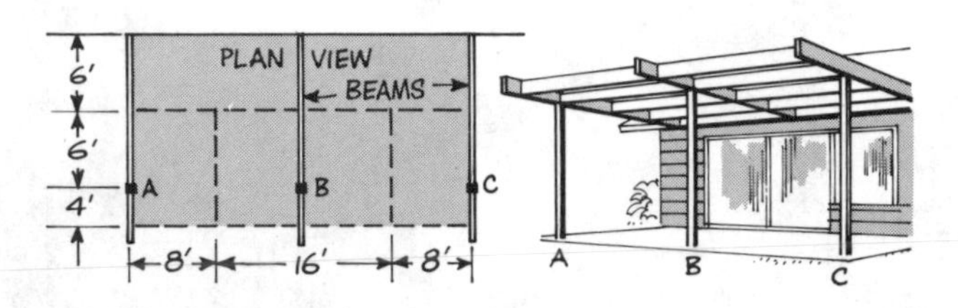

(Example: In the patio roof sketched above, the total load supported by posts A and C is equal to 80 square feet multiplied by 30, or 2,400 pounds. The load supported by post B is equal to 160 square feet multiplied by 30, or 4,800 pounds.)

Don't overlook a number of other post possibilities, such as steel pipe and spaced or built-up posts. Steel pipe is exceptionally strong: a 2-inch wrought steel pipe, 8 feet or less in length, will support 10,000 pounds; a 1½-inch pipe, 5,000 pounds. The primary advantage of the spaced or built-up post is that it offers more imaginative design possibilities than the solid post.

Figuring beam size is somewhat complicated, but the table on page 19 will help on most jobs. Two building inspectors we interviewed suggested this rule of thumb for checking beam sizes: For a 4-foot span, a 4 by 4-inch beam is sufficient; for a 6-foot span, a 4 by 6-inch beam; for an 8-foot span, a 4 by 8-inch beam; and so on. They use the "4-by" (4-inch beam width) as a constant and make the beam depth in inches match the beam length in feet.

The beam sizes listed in the table demonstrate that, though this is a "safe" method of figuring beam sizes, it is not necessarily the most economical, or ideal, way to achieve a delicate structure—especially with short rafter lengths.

Like the rafter sizes, the beam sizes given apply only for those moderate climates where snow load need not be considered. If you live in a colder climate and plan to build a solid, closed-in roof, it would be wise to check with an engineer or with someone in the local building department to determine the necessary dimensions for beams.

Rafters

Rafter sizes are the easiest to determine. You need to know only the length of the rafter and the center-to-center spacing. In moderate climates where snow loads need not be considered, the table on page 19, which is based on temperate Pacific Coast conditions, will give you the necessary rafter sizes. In localities where snow loads must be considered, building departments generally maintain tabulated data, obtainable on request.

BUYING THE LUMBER

Here are some practical ideas that will help you to design your patio roof and order the lumber:

1. Design the structure so that lengths of rafters, beams, and posts are in even numbers of feet where possible. Any even-foot length up to 20 feet is generally available at any lumber yard and can be ordered without delay. Lengths up to 36 feet are usually available on special order.
2. Design with common dimension lumber—odd sizes are not always carried in stock and may have to be ordered. These are considered common sizes (in inches): 2 x 4, 2 x 6, 2 x 8, 2 x 10, 2 x 12, 4 x 4, 4 x 6, 4 x 8, 4 x 10.
3. Design with roofing material in mind—sheet sizes available, widths of roll goods, lengths of lath. Lath, reed, and bamboo usually need closely spaced rafters so they won't sag.
4. When buying lumber, specify quantity, grade, species, size, and length desired, rather than total length and size only. (When ordering, say "12 No. 2 Douglas fir 2 by 4s 20 feet long" instead of "240 feet of 2 x 4 lumber.")

Chart of material requirements

In the chart on page 19 we have confined your choices to standard sizes of lumber and to dimensions that are adapted to most of the patio roofing materials available to you. If you desire to build a patio that does not fit the dimensions listed in the table, check your plans with an engineer. We have arbitrarily set a limit of 16 feet for spanning either with rafters or with a beam because this is the maximum that non-professionals should attempt.

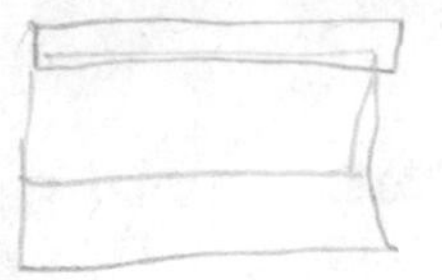

MATERIAL REQUIREMENTS FOR BASIC STRUCTURE

(For a patio roof supported on one or more sides by the house walls.)

PATIO ROOF AREA	RAFTERS—Spacing and sizes	BEAM—Size and number required	NUMBER OF POSTS— 4″ x 4″ x 8′ to 12′
8′ x 16′	12″ o/c—2″ x 4″ x 8′ 16″ o/c—2″ x 4″ x 8′ 24″ o/c—2″ x 6″ x 8′ 32″ o/c—2″ x 6″ x 8′	2—2″ x 8″ x 8′ 2—4″ x 6″ x 8′ 1—2″ x 14″ x 16′ 1—4″ x 10″ x 16′	3— 8′ o/c Same 2—16′ o/c Same
8′ x 20′	Same	2—2″ x 10″ x 10′ 2—4″ x 6″ x 10′	3—10′ o/c Same
8′ x 24′	Same	2—2″ x 10″ x 12′ 2—4″ x 6″ x 12′ 3—2″ x 8″ x 8′ 3—4″ x 6″ x 8′	4— 8′ o/c Same 5— 6′ o/c Same
10′ x 16′	16″ o/c—2″ x 6″ x 10′ 24″ o/c—2″ x 6″ x 10′ 32″ o/c—2″ x 8″ x 10′	2—2″ x 8″ x 8′ 2—4″ x 6″ x 8′ 1—4″ x 10″ x 16′	3— 8′ o/c Same 2—16′ o/c
10′ x 20′	Same	2—2″ x 10″ x 10′ 2—4″ x 8″ x 10′	3—10′ o/c Same
10′ x 24′	Same	2—2″ x 12″ x 12′ 2—4″ x 8″ x 12′	3—12′ o/c Same
12′ x 16′	12″ o/c—2″ x 6″ x 12′ 16″ o/c—2″ x 6″ x 12′ 24″ o/c—2″ x 8″ x 12′ 32″ o/c—2″ x 8″ x 12′	2—2″ x 10″ x 8′ 2—4″ x 6″ x 8′ 1—4″ x 12″ x 16′ Same	3— 8′ o/c Same 2—16′ o/c Same
12′ x 20′	Same	2—2″ x 10″ x 10′ 2—4″ x 8″ x 10′	3—10′ o/c Same
12′ x 24′	Same	2—2″ x 12″ x 12′ 2—4″ x 8″ x 12′	3—12′ o/c 4— 8′ o/c
16′ x 16′	16″ o/c—2″ x 8″ x 16′ 24″ o/c—2″ x 10″ x 16′ 32″ o/c—2″ x 10″ x 16′	2—2″ x 10″ x 8′ 1—4″ x 8″ x 8′ 1—4″ x 14″ x 16′ Same	3— 8′ o/c Same 2—16′ o/c Same
16′ x 20′	Same	2—2″ x 12″ x 10′ 2—4″ x 8″ x 10′	3—10′ o/c Same
16′ x 24′	Same	3—2″ x 10″ x 8′ 3—4″ x 8″ x 8′ 2—2″ x 14″ x 12′ 2—4″ x 10″ x 12′	4— 8′ o/c Same 3—12′ o/c Same

Source for this table is the Uniform Building Code published by the International Conference of Building Officials. Its requirements apply in many communities, but it would be wise to check with your local building inspector to determine whether local standards are more applicable.

HOW TO BUILD IT

Building an overhead is not a difficult or complicated task as construction jobs go, but there are tricks to learn if you want to do it properly. It is easy to make simple but serious mistakes that may cause trouble when the overhead goes into its first stormy winter.

If you have selected your materials from the chart (see above), you should have no worries about failure of the structure from flimsy members. When attaching the structure to the house, assembling rafters and beams, and supporting the posts, however, the following pointers may help you to avoid serious errors.

Connecting to the house

You can attach the patio overhead in one of three ways: 1) to the roof of your house, 2) to the wall, or 3) to the eaves. Which method you select will be largely determined by the height of the eaves above your patio floor level. You naturally want to build your roof high enough so it clears outward-swinging doors and windows and doesn't stun your tall guests. But, if the eaves are low to the ground—as they are on slab-floored houses—and if you are using 6 or 8-inch rafters, you may find yourself forced to anchor your overhead above the eave line in order to gain needed clearance.

Attaching to the wall. If your eave line is high enough to permit your bringing the rafters for the overhead in under the gutter, you can anchor directly to the house wall. This is probably the best way to attach the overhead to the house.

Simplest way to attach to the wall is to fasten a long board to the wall upon which you can rest the rafter ends. This strip is called a "ledger." Attach it securely to the studs with lag screws or spikes. Be sure you fasten only to the studs, otherwise the ledger is apt to give way. If your studs do not reveal their location by rows of nails in the siding, locate them precisely with a stud-finder or by thumping the wall. Since studs are normally set 16 inches apart on center, if you can locate one of them exactly you can find the others with a tape measure. However, don't forget that studs are spaced closer together at corners and around doors and windows.

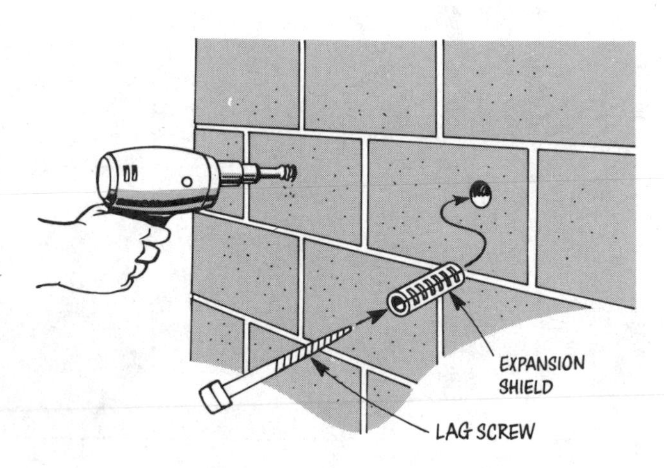

If your wall is stuccoed, you will have to drill holes through it for the spikes or lag bolts. Pound the hole with a star drill or bore it with a masonry bit in a power drill. If the wall is masonry—brick, adobe, concrete blocks—drill at least 2 inches into the wall and then fit lead sleeves into the holes (as shown in the drawing).

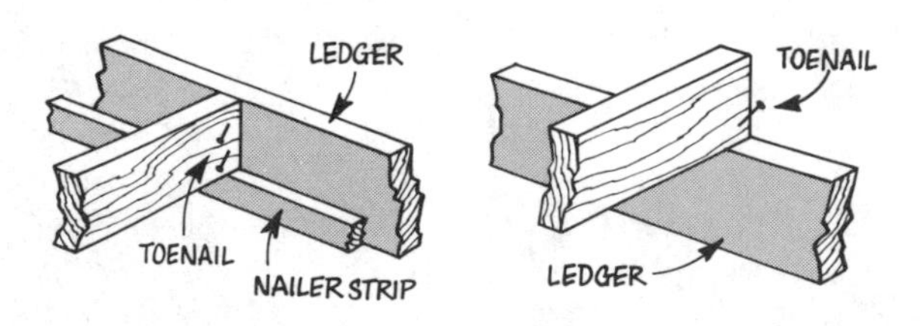

The strongest way to support the rafter ends is to rest them on top of the ledger strip and toenail them to the support. For this installation, a ledger of 2 by 4-inch lumber is adequate.

If the space under the eaves is too cramped to permit the ledger to come below the rafters, you can butt the rafter ends against the strip and secure them with metal hangers.

Attaching to the eaves. Under some circumstances the best way to anchor the patio roof is to attach it directly to the eaves or to the fascia (facing strip) already attached to the edge of the roof.

One word of caution is in order: the eave line is the weakest possible point for attaching additional weight to many roofs. In some forms of roof construction, the rafters extending beyond the house are only capable of supporting the roof overhang. Any heavy load added to them would cause them to sag. Because the fascia on many roofs is purely ornamental, it too is not designed to support any weight beyond its own.

As a rough rule of thumb, you can figure that 2 by 4-inch roof rafters will support one side of a

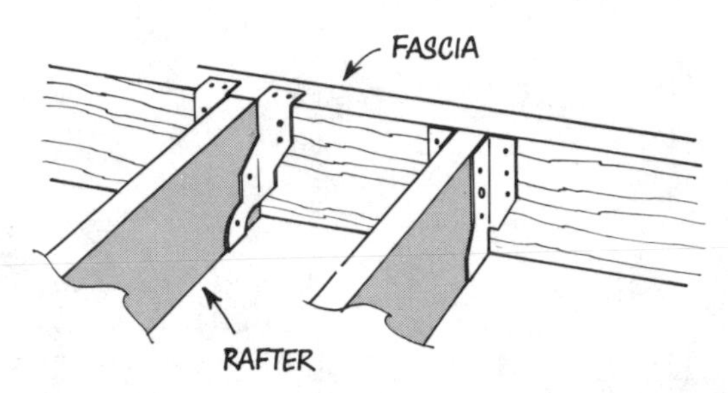

patio roof if they do not extend more than 24 inches beyond the house; 4 by 6 or 4 by 8 rafters on a flat roof will serve if they do not extend beyond 4 feet. If your roof's overhang is greater than these limits, check your problem with an engineer or your building inspector.

As for the fascia, examine it closely before you attach weight to it. Most likely, it is simply nailed to the tips of the roof rafters. If this is so, you will have to supplement the fastenings with metal strips (as shown in the sketch). If the fascia is 1-inch

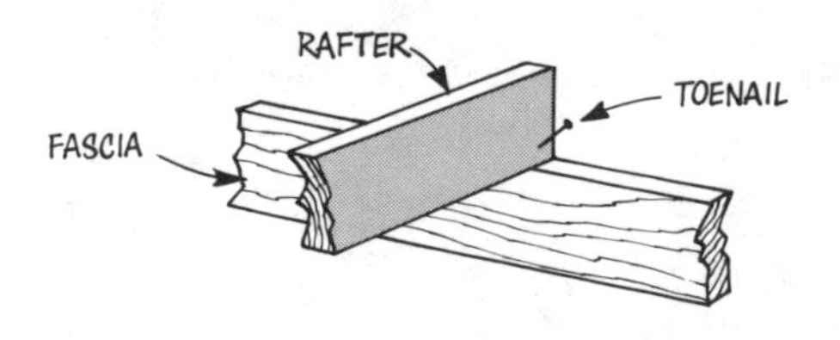

stock, add another piece of 1-inch stock to double the thickness for the distance that it supports the patio roof or else replace it with a 2-inch board.

Attaching to the roof. Anchoring the overhead to the roof above the eave line is a good way to solve one or both of two problems: it provides head room when the eaves are too low, and it permits free escape of warm air from a closed-in patio.

The standard method for attaching a patio roof in this fashion is to lay a ledger strip right on the shingles or composition roof and attach it with lag bolts run through the roofing to the roof rafters. Holes made in the roof for the lag bolts are sealed with mastic to prevent leakage. The patio roof rafters are then toenailed to the ledger, as in the drawing.

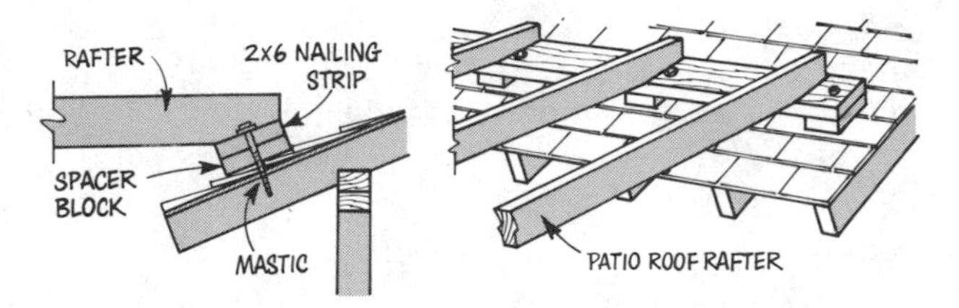

The ideal point of attachment for the ledger strip is directly above the wall of the house because this will distribute the extra load down the wall without straining the roof in any way. However, since this type of attachment is difficult to make watertight—rain water will seep in around the holes after the mastic dries out—it is recommended that the ledger be positioned just forward of the wall line. If any leakage does occur, it will not damage interior woodwork.

Installing rafters

Rafters in the patio roof have a double task to perform: they must support their own weight over open space without sagging or twisting, and they must also support the added weight of the roof covering. This dual capacity explains why the rafter dimensions recommended in the chart may seem excessive. The familiar 2 by 4, for instance, does a fine job in house framing where it stands vertically and is well braced; but if it is used horizontally as a rafter, it is subject to failure.

The proper size of rafters for your patio overhead is given in chart on page 19.

Blocking and cross-bracing. Rafters that are sturdy enough to support their own weight as well as that of the membrane above are subject to a special weakness that you will have to guard against. Prone to twisting, they require bracing at each end and sometimes in the middle to keep them in line. The rule of thumb: if the rafter's ratio of depth to breadth is 3 to 1 or more, it requires cross-bracing at each end. Thus, a 2 by 6 (3:1 ratio) or a 2 by 8 (4:1 ratio) needs blocking, but a 4 by 8 (2:1 ratio) does not.

Blocking used for this purpose is usually as substantial as the rafter timber itself. A series of 2 by 6 or 2 by 8-inch blocks is more than adequate to hold 2 by 6s and 2 by 8s in place (see sketch

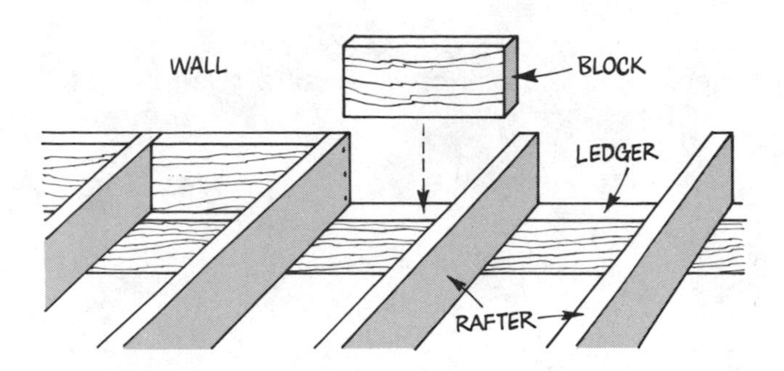

above). One way to economize is to substitute 2 by 4-inch blocks aligned with the tops of the rafters.

How to estimate overhang. An overhang of at least a few inches pleases our senses more than an abrupt drop-off. But there is a practical limit to the amount of overhang that you can build with safety. As you can see from the following table, the greater the overhang, the heavier the structure has to become in order to compensate for the shifted load. If the options offered in this table do not meet your requirements, you should refer your design to an engineer for development.

RAFTER SPACING CENTER TO CENTER	LUMBER DIMENSIONS			
	2 x 4	2 x 6	2 x 8	2 x 10
16"	2'0"	3'0"	4'0"	5'2"
24"	1'8"	2'8"	3'6"	4'6"
32"	1'6"	2'6"	3'2"	4'0"

How to install pitched rafters. Patio cover materials that shed water must be pitched. The slope depends on the covering material—from 0″ to 1″ per foot for built-up roofs to 4″ per foot for wood shingles and shakes, and asphalt shingles. If the patio roof is above the eave line and if it is not too large in area, it can be pitched toward the house roof so its rain runoff will be carried away by the house gutters.

Fitting sloping rafters in place is an exacting operation. Carpenters follow many different ways of doing this, but the amateur is probably most likely to do a good job if he follows the old-fashioned method of cutting one rafter to fit and then using it as a template for the rest. Here is the procedure:

1. After the ledger strip, posts, and beam are in place, lay a rafter board so it rests on edge on both beam and ledger.

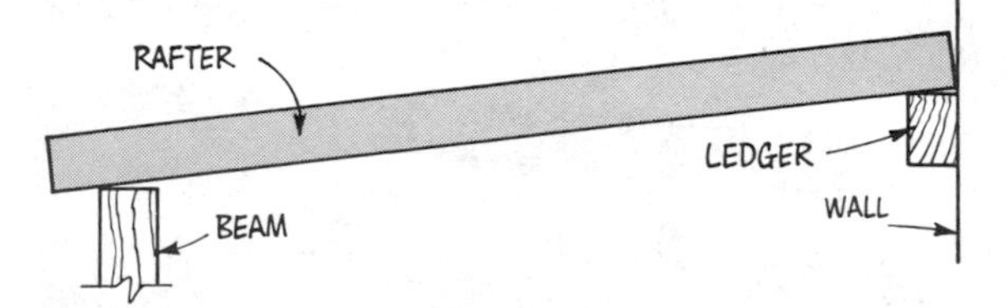

2. Force the tip snugly against the house wall, then, using a block of wood as a ruler, mark the end for cutting.

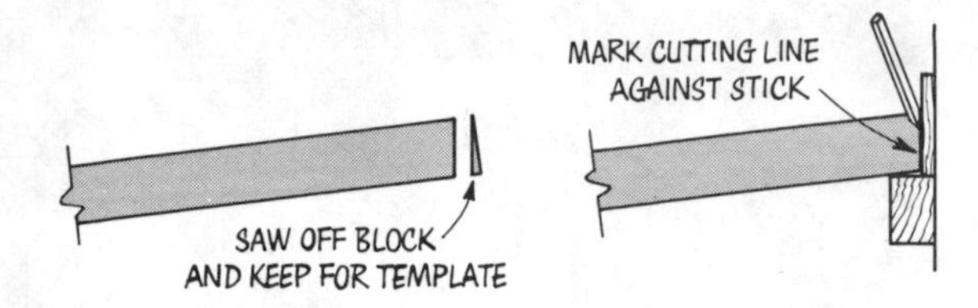

3. Cut the triangular piece off the rafter end where it rests on the ledger strip and on the beam, as shown.

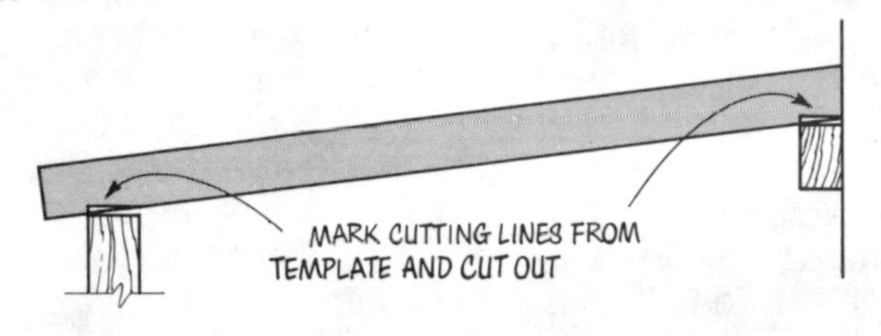

4. Cut out the notches you have marked and try the cut rafter in place to see if it fits, and also check the fit in several other positions.

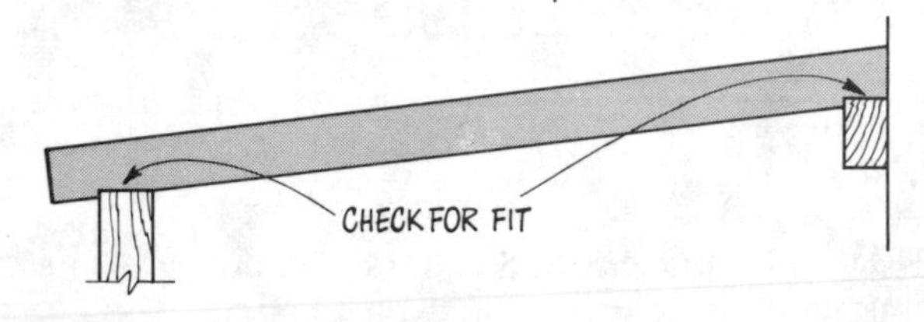

5. Now, using the first rafter as a template, mark the remainder of the rafters for cutting.

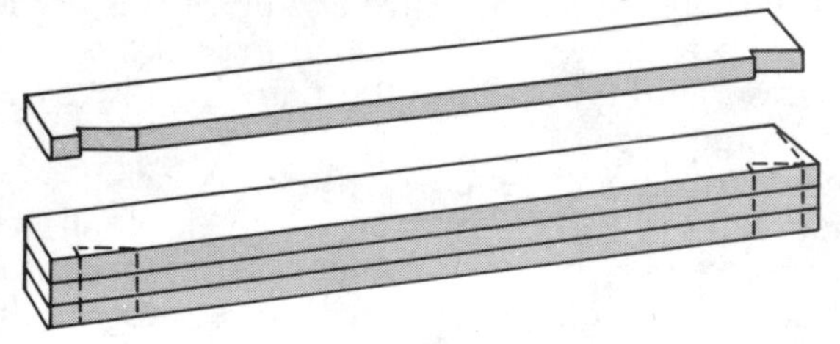

6. Toenail them in place. If you wish to do a deluxe job, slosh some sealer-preservative on the two mating surfaces of the joint before nailing in place.

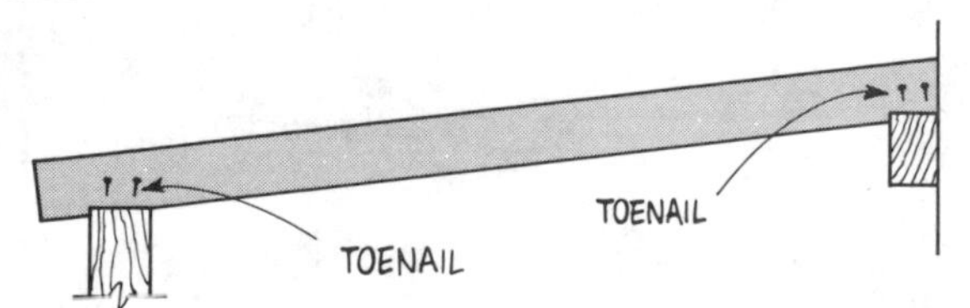

Installing beams

The beam (some call this a lintel) takes the forward part of the roof load and transfers it to posts on its way to the ground. The correct size of beam to use is determined by the weight of roof it must carry and the distance it must span. Recommended dimensions are listed in the chart on page 19. If these do not fit your needs, check with an engineer.

How to attach to posts. You can choose from many different ways of attaching the beam to the posts, but the strongest method is to rest the beam on top of the post, and fasten it in place with straps or by toenailing.

How to support against the house. When a patio roof is built over a terrace walled in on two or three sides by the house, one or both ends of the beam must be supported against the house. The temptation is to use the house wall as a support, but you should avoid this unless you are willing to open up the house wall and attach the beam directly to the exposed studs. Even then, you should run a support down inside the wall to the plate. Actually, the load carried by the beam ends is too concentrated to entrust to the house wall, and the beam should be given support that is independent of the house framing, although it may be tied in for added strength.

The safest method is to provide a separate post for the beam, set out from the wall and resting on its own footing.

Next best is to bolt a 4 by 4 or a steel-capped 2 by 4 to the studs and run it down to its own footing, poured alongside the house foundation.

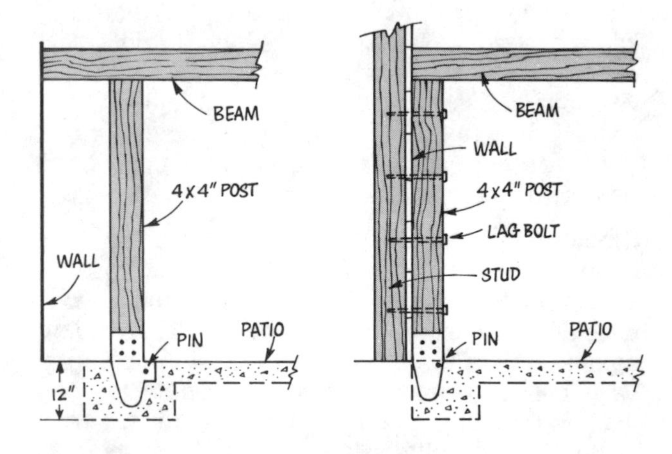

How do you get it up there? One simple fact that you might overlook in your calculations is that the beam you plan to install may weigh from 200 to 500 pounds (average 50 pounds per cubic foot). Green wood may weigh much more than this.

There is no sure-fire method for lifting the beam into position—aside from inviting your most muscular friends over for a Saturday afternoon—but you can try this suggested routine:

1. Drag the beam into rough position on the ground.
2. Lift one end and put a 2 by 4 under it.
3. Nail a cleat firmly on each side of the top of one post to act as a cradle for one end of the beam and to keep it from slipping off.
4. With the help of your assistant, lift the beam end with the 2 by 4 and wiggle it into the cradle on top of the post.
5. As a safety factor, you can cinch the beam to the post with rope, but allow enough slack to let the beam slide forward far enough to rest on top of the next post.
6. Now get under the other end with the lifting bar and heave it up to the top of the second post sliding it into position.

If you have access to a block and tackle and have the know-how for using it, you can also save yourself a possible backache.

Erecting posts

As stated above, the 4-by-4 post is standard for all patio overheads. You are strongly advised against using smaller-dimensioned lumber, such as the 2 by 4, because it may bow as it adjusts to the load. Furthermore, small posts cannot carry as much weight. However, paired 2 by 4s, nailed together as shown in the drawing, will substitute nicely for 4 by 4s; in fact, under some circumstances, the paired 2 by 4s can provide better support than the single 4 by 4 because they spread the load. Spacer blocks should be used between the paired 2 by 4s.

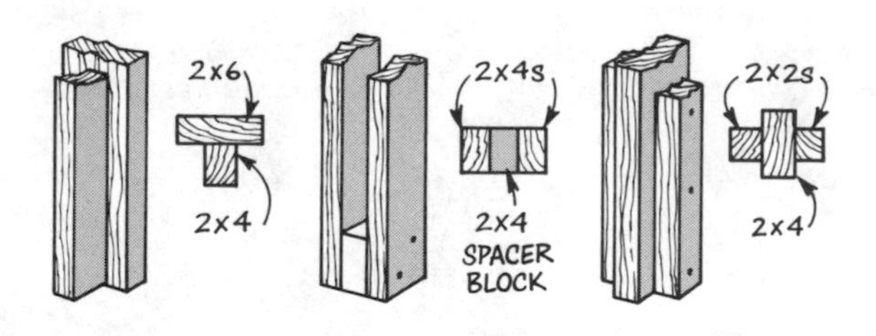

Footing requirements. Posts should always rest on a concrete foundation or pier. Although a 4-inch slab of concrete might be sufficient to support a post carrying a lightweight or open roof, a pier or foundation 14 inches deep and 12 inches wide will be needed if the outdoor room is ever to be roofed conventionally or closed in.

Note one fact, however: unlike fence posts, which stand rigid as soon as their concrete footings have set around them, these posts are wobbly and unruly until they are load-bearing because their ground connections depend on the roof load for part of their rigidity. The pin insert is particularly vulnerable—if the posts fall over, they bend the center pin out of line. So you will have to brace them with wire, rope, or wood outriggers while you fit the beam and rafters into position.

Bracing

Four posts and any roof, or even two posts and a roof connected to one wall of a house, need one more assist to be structurally sound. They must be braced to resist lateral forces—primarily the wind.

You gain "built-in" bracing if you build the roof in the lee of two house walls or make it an egg-crate structure or a solid sheathed diaphragm firmly attached to the house.

You need actual bracing at the posts if your structure is merely an open one consisting only of beams and rafters.

The most common solution, a simple knee brace at the posts, may be the most ungainly, especially if large-dimension wood is used.

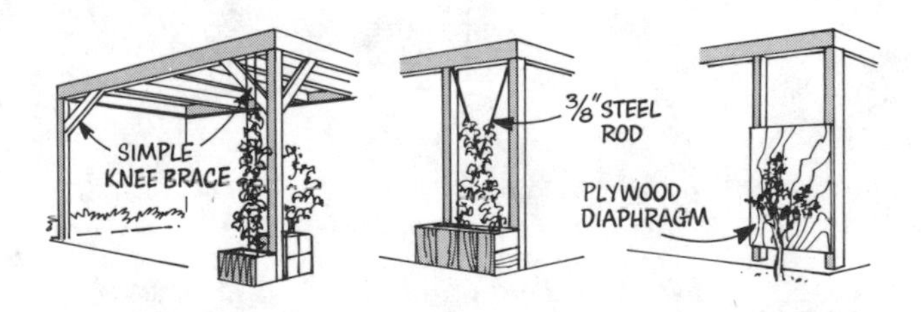

By changing the shape and character of this bracing, however, you can often improve the appearance of the structure.

How about steel frames?

For strength, endurance and a trim appearance, designers frequently specify steel as a construction component for the roof's supporting framework.

This material has great advantages—as well as disadvantages. In its favor, steel is a clean and graceful material that can be used to produce a garden structure of delicacy and beauty. The clumsy 4-inch wood post, for instance, is replaced by the 2-inch, or even 1½-inch pipe. Heavy wooden beams are displaced by steel I-beams half their size, and the standard 2 by 6 rafter by thin channel iron.

(Continued on page 25)

HOW TO CONSTRUCT A PATIO OVERHEAD

Building an overhead is not a difficult or complicated task as construction jobs go, but there are points to learn if you want to do it properly. Here is a suggested sequence to follow to help eliminate the possibility of making any serious mistakes.

In addition to hammer, saw, and stepladder, you will need a device for establishing level lines. A line-level and 50 feet of chalk line will do nicely, although a carpenter's level can also be used with a long straight board. A 50 or 100-foot tape measure is useful.

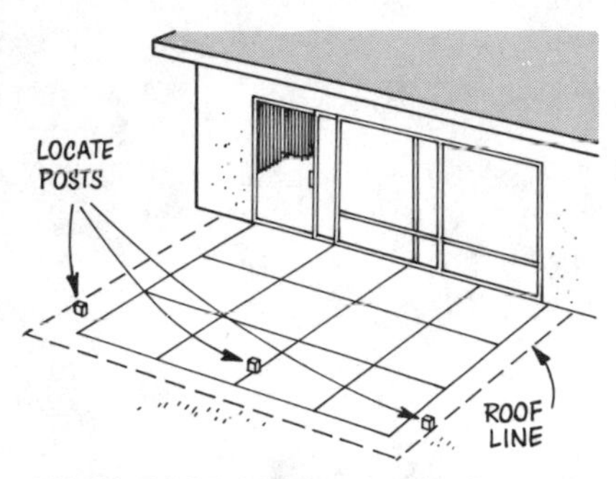

MARK ROOF LINE on patio floor with stakes and string. If the roof will be pitched to shed water, don't forget to make allowance for this. Then mark the location of the support posts.

PREPARE POST FOOTINGS in one of three ways: in bare ground, dig holes 14 inches deep, 12 inches across; on thick slab, drill hole for metal pin to attach to post; for weak slab, use sledge and dig.

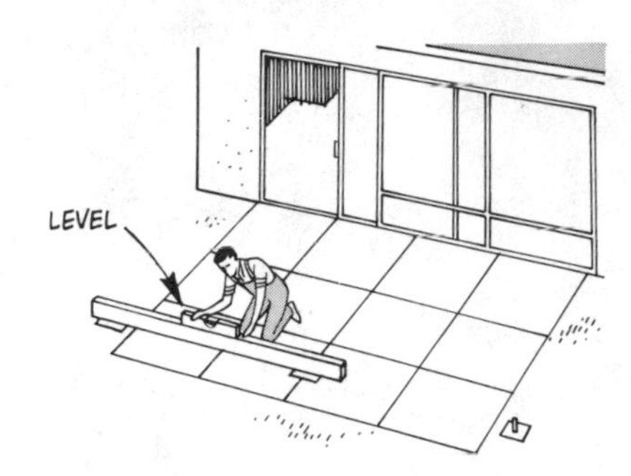

FILL HOLES with concrete and set in anchor pins or brackets needed to hold the posts erect. Use a line level or carpenter's level placed on a straight board to check the level surface of each footing.

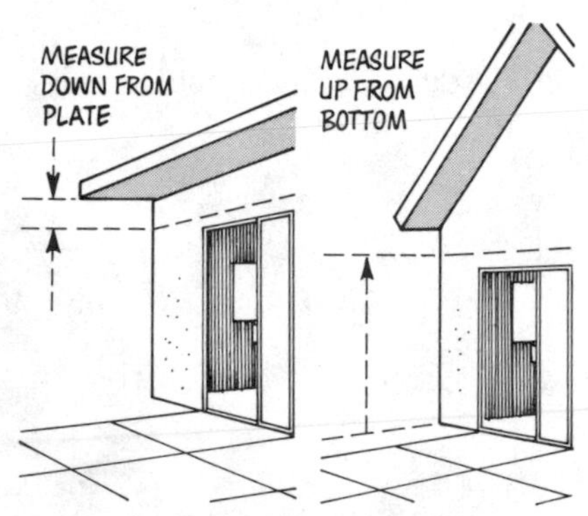

MARK TOP ROOF LINE on house wall measuring (a) down for a patio roof that will run below eaves of the house and (b) up from the foundation for a patio roof to be attached to a pitched wall.

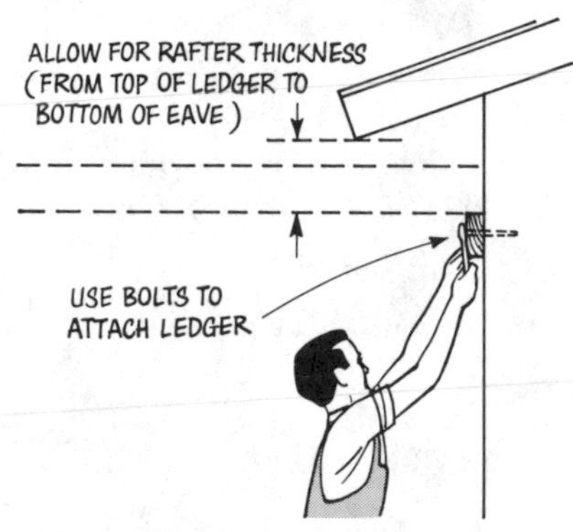

ATTACH LEDGER STRIP to house studs with lag bolts or screws. If the wall is stuccoed, first drill holes with a star drill; if the wall is masonry, drill at least 2 inches into wall and fit in lead sleeves.

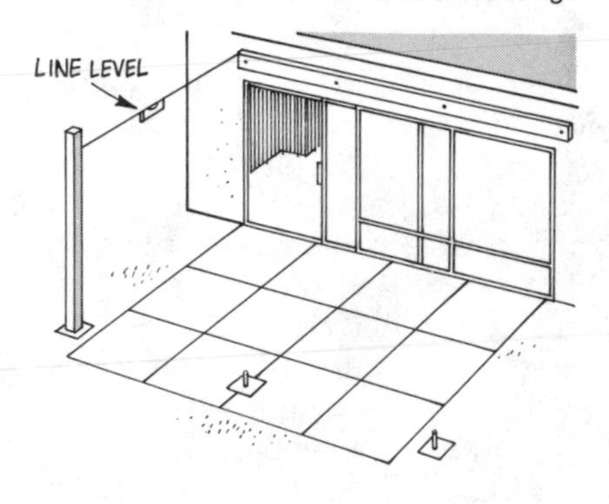

MATCH POST HEIGHTS to height of ledger strip in this manner: set a post in place and square it up, then run a line-level from top of strip to post, level the line, and mark the post at the line.

MEASURE DOWN from the rafter line a distance equal to the thickness of the cross beam and mark a cutting line. Cut off the end of the post at this point. If footings are level, cut other posts.

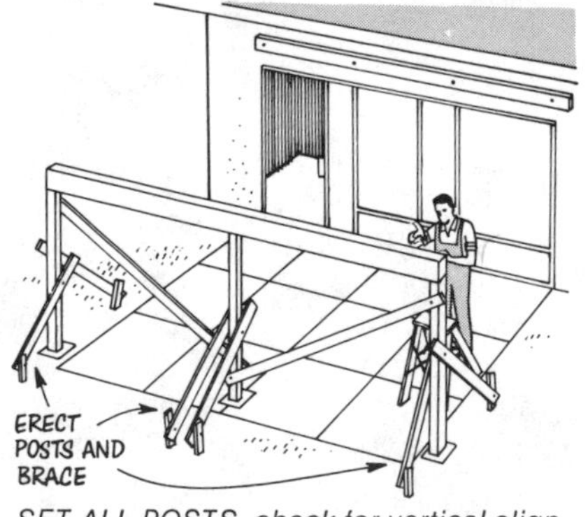

SET ALL POSTS, check for vertical alignment, and brace each one with stray pieces of lumber. Then hoist the cross beam on top of the posts, level, and fasten it securely in place.

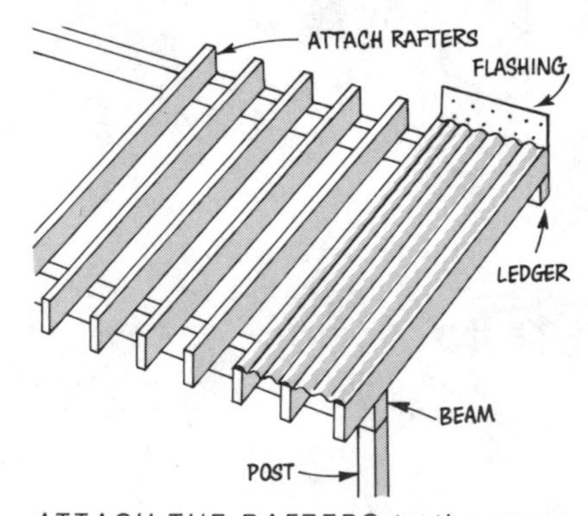

ATTACH THE RAFTERS to the cross beam and to the ledger by using one of the methods illustrated on page 17. Use blocking for cross support if necessary. Attach patio covering; needed flashing.

(Continued from page 23)

On the other hand, steel is expensive, difficult to obtain in small lots, and well beyond the craft skill of anyone but a steelworker. In addition, it requires annual repainting to protect it from rust.

However, there are two areas where the amateur can feel at home with the material; the simple pipe frames used to support canvas, reed, and bamboo—as described in later chapters—are easily assembled; and the pipe post, substituting for the usual 4 by 4, can also be mastered.

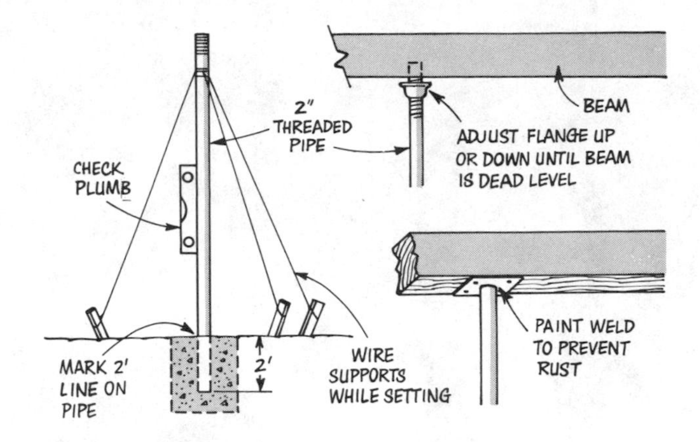

How to install pipe post. Use of 2-inch iron pipe for a post support calls for one special warning. Steel is a remorselessly precise material to work with. The post has to be exactly vertical—unlike wood, it cannot be nudged back into place. It must be cut to the exact length, because you can't shim it out with strips of wood nor can you cut it shorter when it is fixed in place. With these warnings in mind, let's check through the procedure:

1. You have two choices: you can either imbed posts in a collar of concrete as shown in the drawing (upper right), or you can thread one end into a flange fixed to the concrete. If you set the post in concrete, imbed it 2 feet, check its plumbness with a level, and support it firmly with wire or rope while the concrete sets. The pipe should be brushed inside and out with a good quality white lead paint before it is set in the wet concrete.

If you plan to use a floor flange, cast the footing for it so it is absolutely level, otherwise the post will stand crooked. When the concrete has set, drill four holes to receive lead shields and lag bolts.

2. The top of the post, where it joins the wood beam, can also be treated in two ways. You can have a plate with screw holes welded to it, or you can thread the pipe into a flange. Have the pipe threaded for about 1½ inches, and drill a 2-inch hole in the beam where the pipe fits. By rotating the flange up or down the pipe, you can compensate for any irregularities in figuring, and can position the beam dead level.

If you have a flange welded to the pipe, paint it liberally, because the protective zinc coating will have been burned away at this point.

THE BASIC FRAMEWORK

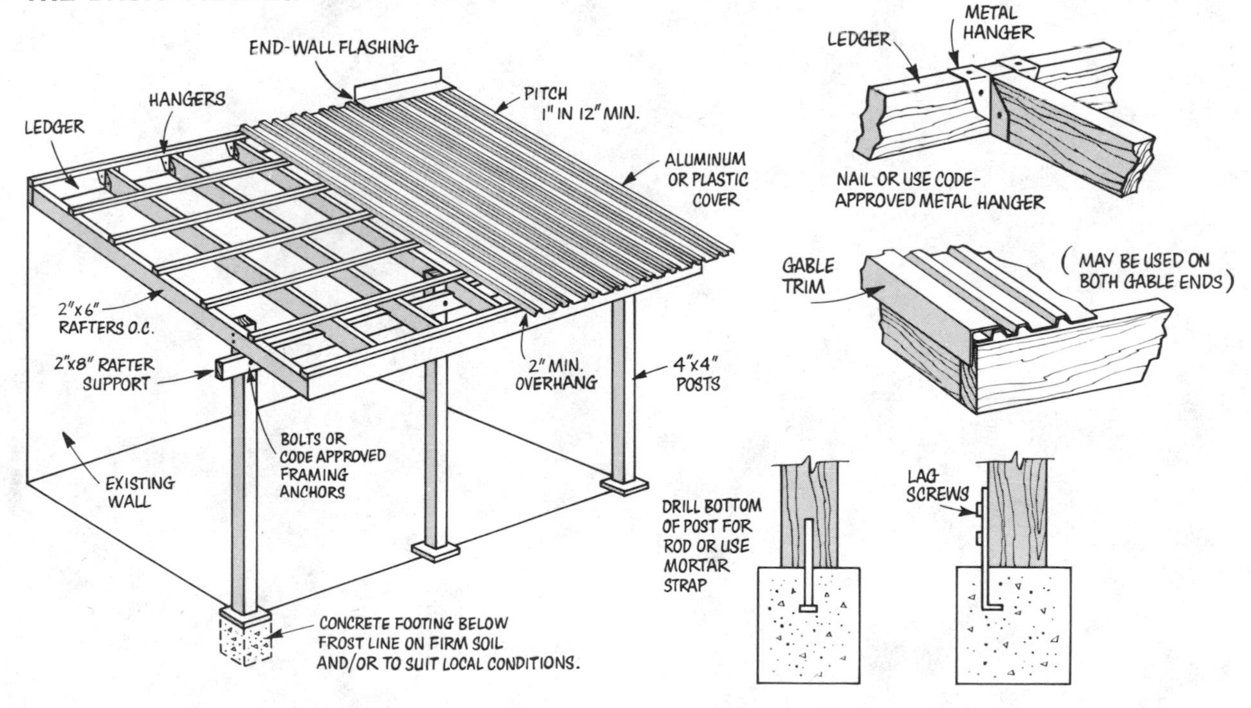

Patterns in reed and bamboo

From Polynesia, the Orient, and the West

GOOD PLANNING, simple materials build authentic teahouse.

Reed, bamboo and the idea of woven woods are natives of the South Pacific and the Orient. With a little semi-controlled daydreaming plus a trip to the neighborhood building supply house, you can create a mini-polynesia right in your own back yard. Or, you may prefer to experiment with the domestic potential of reed and bamboo, either as removable frames or permanently nailed down.

THE MATERIALS

Among the lightest and perhaps the most interestingly textured overhead cover materials are woven bamboo and reed. The fact that both materials are good shade makers, inexpensive, and easy to handle helps to account for their wide use, but their handsome appearance certainly accounts for much of their popularity.

These materials have a lot in common. Both cast soft, irregular shade patterns that many people prefer to the sharply defined striped shadows cast by lath overheads, both require a minimum understructure because of their light weight, and either one is easier to install than almost any other material.

Lightweight reed rolls

Woven reed comes in 15 and 25-foot rolls 6 feet wide and is woven with a stainless steel wire to withstand the attack of weather. The wire can be easily cut and retwisted when the roll is being trimmed to the dimensions of the overhead, but the stiffness of the wire itself keeps the woven reed from being freely adjustable in an overhead screen. Constant flexing of the wire strands causes them to fail quickly. If the woven reed is nailed or stapled to a rigid frame, however, there is no problem.

BAMBOO shades patio overlooking a tropical valley on the island of Oahu. (Landscape Architects: James Hubbard and Paul Weissich.)

The reed rolls are available at many nurseries and garden supply stores and can be counted on for several seasons' use.

If you make provisions for easy removal so that the reed cover can be stored indoors in the winter, the life span may be immeasurably increased. To make it easy to remove, either nail or staple the reed to removable panels or secure the reed to the framework with 1 x 2-inch cleats that can be removed easily with a claw hammer.

Some builders use the reed overhead and repeat the material in sections of fencing or interior windbreaks. This is another use the reed is quite suited for, provided it is secured to a rigid frame.

The reed can span its full width with a single support down the middle without noticeable sag. Each span, then, would be 3 feet 2 inches and the length would be limited only by the maximum roll length—25 feet.

Because the material is so light, don't be tempted to put up an understructure that is too flimsy—see the chapter on construction of the supporting structures page 15.

Reed, like bamboo, can be laid on wires stretched tightly between two secure anchors; but unlike bamboo, the reeds cannot be quickly rolled up to escape stormy weather. You have to take down the whole screen.

A stapling gun similar to the one that is used to install window screening will speed installation if you are permanently attaching the reed to frames. Be sure you get weather-resistant staples long enough to straddle the individual reeds. Common 1-inch galvanized staples will also do the job.

Split and matchstick bamboo

Woven bamboo, which is manufactured primarily for vertical shade use, comes in rolls of varying widths from 3 to 12 feet and a standard length of 6 feet.

There are two main grades: split and matchstick. The split bamboo is coarser and less regular than matchstick, which is made from thin strips of the inner layer of the bamboo stalk. It is available in wired form, as is reed. Price will vary depending upon the quality.

Split bamboo, stiffer than matchstick, is preferable for most installations. For an adjustable overhead suspended from wires, however, the matchstick is often preferable because of its flexibility.

Bamboo may be dipped in a water-repellent preservative to insure maximum lasting quality, but this can be an exacting task. Because bamboo is relatively inexpensive, many people forego the initial treatment, preferring to replace the material whenever it begins to show excessive signs of wear.

TWO METHODS FOR ROOFING

Although alternate methods of applying reed, bamboo and woven wood coverings exist, two techniques seem to have won general acceptance. Adjustable covers and box beams afford simple lightweight construction, economy and ease of handling.

Woven spruce or basswood

Another material that is similar to bamboo but has a more polished appearance is woven spruce or basswood shade. Like bamboo, it is woven with string, but because the wood shade is made specifically for outdoor use, a high grade of seine twine is used. This twine—used in fishing nets—will probably last as long as any cotton cord.

Treatment with water-repellent preservative every three years will assure maximum longevity.

How to make box beams

Adding box beam extensions to an existing roof overhang is an excellent way to provide support for reed, bamboo, or woven wood (see photographs on the facing page). It keeps low winter and spring sun from building up heat in the house and breaks the driving force of rain. The most interesting thing about it is that it extends 6 feet beyond the roofing without posts.

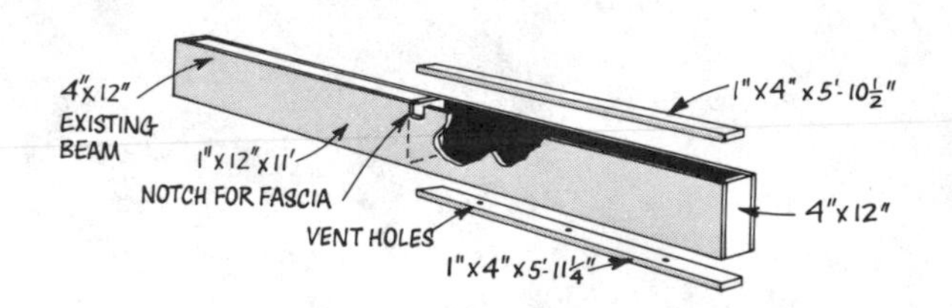

You can prefabricate these beams right on the ground. The two 1 by 12s on edge provide great supporting strength without having the weight of a solid beam. No pitch is necessary for water drainage, since the screen is not a solid material.

Several variables affect the length and load of such a cantilever. Be sure to obtain the approval of your building inspector; it would also be wise to seek the advice of an architect or engineer.

When cutting out the parts, be sure to notch the 11-foot side lengths on their upper edges to receive the fascia.

After gluing joints with waterproof glue, fasten 1 by 4s with five-penny galvanized nails, 6 inches on center. Fasten 1 by 12 sides of each box beam to the existing beam sides with 8-penny nails, also 6 inches apart. Putty the joints, then paint the beams to match the exterior trim.

1.

2.

How to make box beams

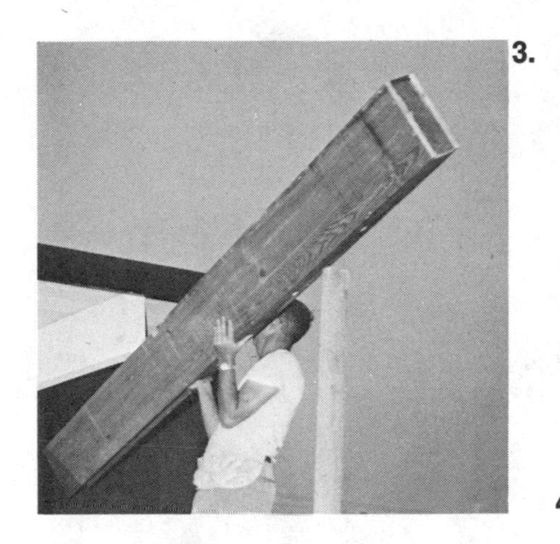

3.

4.

__1.__ TO MAKE BOX BEAM SUPPORT glue and nail members together as you would a box. Diagonally laid boards serve as temporary dividers and spacers. __2.__ Bore ventilating holes in the bottom 1 by 4 in order to prevent rot in hollow beam. Next, staple screening on the inside over holes to keep out insects. __3.__ While helper holds a prop ready, slide the open end over the entire exposed roof beam. Prop firmly, then nail box beam to both sides of existing beam. __4.__ Staple reed along first beam, then pull it taut and staple it to others. Nail 1 by 2-inch bracing "strongbacks" to beams; staple reed to bracing from below. __5.__ Finished product: lightweight box beams and reed screen add 6 feet to existing roof overhang over south-facing glass wall for total of 11 feet of width. No posts obstruct view or interfere with patio activities.

5.

Adjustable covers you can make

If you plan to make an adjustable cover, keep several things in mind:

1. Both the bamboo and the wire will be affected by the weather, so if you can plan an easy way to remove the wires (and thus the shades) you can store them inside during the wet months. If you use turnbuckles, which will make it easy to cinch the wires up tightly, you have no problem. Use snap fasteners, or the lead barrels with set screws that normally come with clothesline wire, if you don't care to bother with turnbuckles.

2. Make sure the wires are firmly anchored. The wind will buffet the bamboo, and its force will be distributed against each of the rings which hold the bamboo to the wire; but the sum of that force will be pulling against the attaching points.

3. Space your rows of rings (which, incidentally, should be made of brass or other weather-resistant material) close enough together to avoid objectionable sagging. Because this material varies so, no single formula could tell how to space. But you can easily unroll the bamboo and lay it across two 2 by 4s, for instance, and see how much it sags. You can then decide how closely to space the rings. Better allow for additional sagging after the material has been in the weather.

FOUR 6-FOOT SECTIONS of matchstick bamboo are suspended on weather-resistant wires.

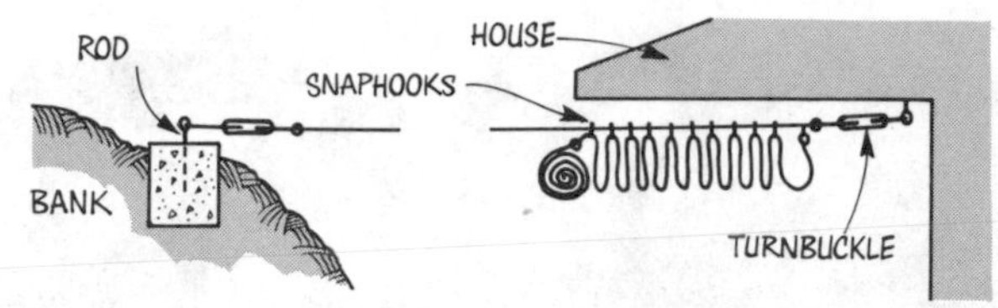

DIAGRAM shows wires attached to bank and house.

FIRMLY SECURED to house and to concrete anchors in hillside, two wires support each roll of bamboo. Although bamboo gives positive shade, patio does not seem to be closed in.

LEFT: pavilion recreation area utilizes bamboo, eggcrate, 6 by 6 posts, heavy duty metal deck anchors (Landscape Architect: James Hubbard). ABOVE: redwood overhead, reed fence, and bamboo gate combine harmoniously in Japanese garden.

Ideas for reed and bamboo

RIGHT: reed covered overhead and windbreak shelter deck that overlooks swimming pool. Diving platform (out of picture) is at right.

Screens of fabric and metal

Sometimes they're the only defense

DECK on the sunny side screens out heat, glare, and insects.

Woven strands—cotton, copper, plastic, steel, aluminum, and bronze—nailed to frames or buoyed into the air on delicate wires give a lacy, filtered light for cool relaxation out of doors.

A wide variety of materials of this type is available to the outdoor builder: he can provide for a permanent dense shade with closely woven netting, merely define the patio roof with an open steel mesh, control the sun's rays through the day with miniature metal louvers, or use a net or mesh in combination with other materials to give a variety of climates in the patio beneath.

Normally, the fine screens are used only when the entire patio is enclosed—either with screen or solid walls.

FABRIC MESHES, METAL SCREENS

When you have decided on what you expect your screen to accomplish, the next step is fairly simple. You can pick out the screen that is best suited to that job. The local hardware or building supply store is usually a valuable source of information and help when you are making your selection. Although the retailer may not stock all of the mesh and screen materials discussed in this chapter, he will, if you ask him, either special-order the material or tell you where it is available.

Flyscreen, first line of defense

Common flyscreen makes an ideal covering for a patio roof where an overhead is desired that does not block out the open sky, where sun control is unimportant—and especially where insects are a summer scourge.

When installed overhead, flyscreen gives the illusion of a free and open sky. Although it somewhat reduces air movement, the reduction is not sufficient to rule it out except in the most severe climates, where the faintest breath of moving air

METAL FRAMED FIBERGLASS shelter keeps coolness in, bugs out.

is treasured. Flyscreen also provides a slight reduction in the sun's glare, for a square foot of metal screen actually covers about 25 to 30 percent of that area with metal.

If a screened roof is paired with screened or solid walls, it provides security from such insect pests as gnats, house flies, and yellow jackets. In some developed sections of the desert where gnats are a widespread nuisance, screened patios make good sense—the wire mesh admits plenty of air but excludes the bothersome insects.

If you decide to have a screen roof, remember that it will collect leaves and twigs and will need an occasional sweeping or hosing off to keep it looking neat and to prevent the screen from rusting prematurely. Unless you are a stickler for neatness, though, the spotted shade pattern made by sunlight filtering through the leaves can be rather pleasant.

Standard insect screen has an 18 by 14 (18 openings per inch horizontally, 14 vertically) and 18 by 16 mesh. It is usually manufactured in 100-foot rolls, the widths of which vary with the manufacturer. Two common size ranges are these: 24 to 36-inch widths in increments of 2 inches, plus 42 and 48-inch widths; and 16 to 48-inch widths in increments of 2 inches plus 60, 66, and 72-inch widths. Sizes over 48 inches must usually be special-ordered by the retailer.

Which screen materials are best?

Here is a list of the major types of screening available in hardware stores or wire mesh outlets (your hardware dealer can direct you to the nearest of these). Other types of screen cloth exist but are less useful for patio screening.

Aluminum. Corrosion resistant aluminum builds up its own protective coating of aluminum oxide and therefore needs no varnish or paint. The aluminum oxide coating, unlike some other metallic oxides, will not cause stains on a light-colored framework. Aluminum screening has a long life expectancy under normal conditions; however, like other metals, it will deteriorate in coastal or industrial atmospheres, though much more slowly. While not as strong as galvanized steel, it tends to bulge if struck or strained rather than to break, as steel will do. In weight it is the lightest of metals.

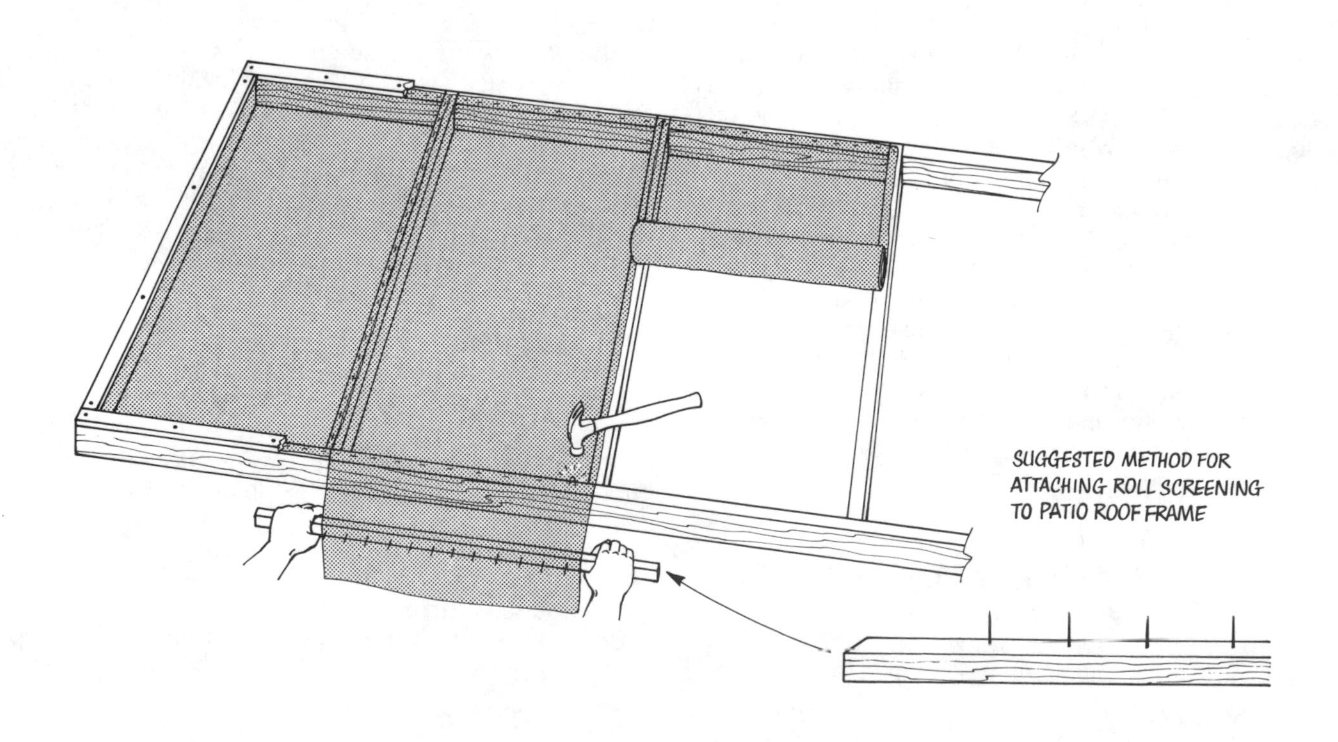

Price range is somewhere between that of bronze and galvanized steel screening.

Bronze. Subject to comparatively little deterioration, but until oxidation dulls it, bronze may prove too bright and highly colored for some large areas. Rain running from it can permanently discolor a paint finish; also, the screen may turn green if soap, acid, or salt in the air comes into contact with it. Varnish applied at least once a year will keep staining to a minimum (do not use paint as it won't adhere to the bronze).

Galvanized steel is the strongest of all metal screening and also the least expensive. It will not last as long as other types of screen because of its tendency to rust, particularly in coastal areas and locations with high humidity. While the zinc coating will prevent rust for a short time, periodic painting is the only sure protection.

Glass fiber. Yarn strands, composed of about 400 extremely fine glass filaments are coated with vinyl (in a range of colors) woven into screen, and heat-set to bond each intersection. Glass fiber screen cannot oxidize or corrode, and the vinyl discolors very little. It is stronger than metal and light in weight.

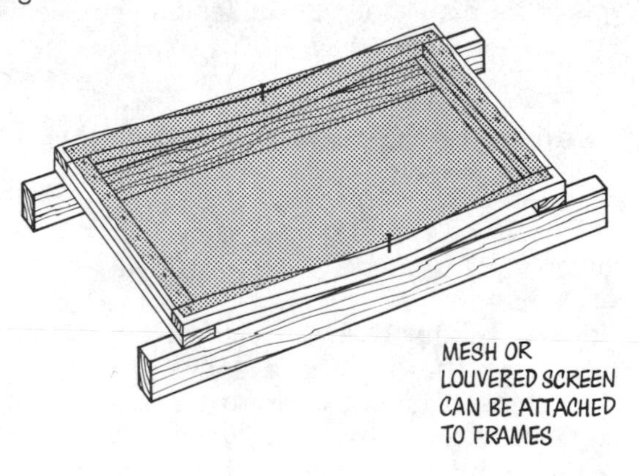

MESH OR LOUVERED SCREEN CAN BE ATTACHED TO FRAMES

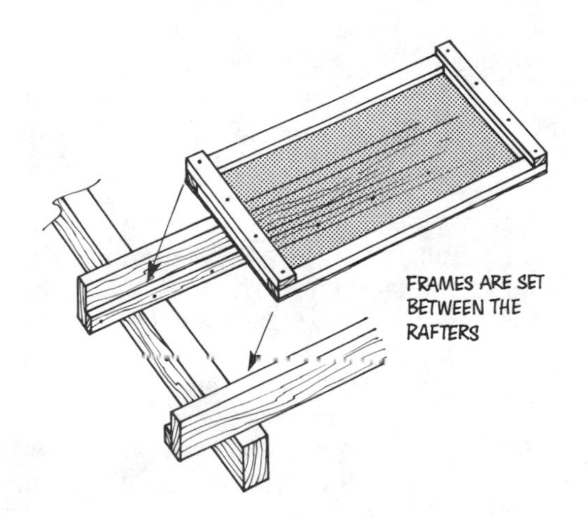

FRAMES ARE SET BETWEEN THE RAFTERS

Since glass screening will not stretch, it can cover larger open panels than other types. Whereas the vinyl will char in flame, the glass will not burn. To repair it you can fuse a patch in place with an iron. It comes in widths up to 84 inches and costs about the same as the more expensive kinds of metal screening.

Plastic. One of the best screening materials for houses located directly on the beach, where the problem is salt air, the plastic mesh will not corrode and it is unaffected by humidity or salt air.

All of these screens are sold by the running foot in widths of 24, 30, 36, 42, and 48 inches; some in a 6-foot width. They can be ordered by the piece (more expensive) or by the roll.

The width you buy will largely be determined by the dimensions of the framework you build, principally the distance between rafters. Although you may be tempted to install a broad width—say, 4 or 6 feet—you might find it difficult to install without sagging. The wide screen is hard to pull taut and awkward to nail. On the other hand, if you select 24-inch screening and tack it to rafters spaced 24 inches on center, you will have to climb up the ladder twice as many times, but you should be able to produce a neater, tighter job, and your close-spaced rafters will fit any one of several other types of patio roofing to which you might want to convert when the screen wears out.

How long can you expect the screen to survive? The answer depends on how damp or dry your climate is and how laden with industrial impurities your air may be. Under severe conditions, screening may deteriorate in two or three years; under ideal conditions, it may last eight or 10 years.

Aluminum-and-plastic. A new introduction, it consists of plastic-coated aluminum wires. The horizontal wires are broad and flat; this reduces sun penetration as much as three-fourths (depending on angle), giving noticeable temperature reduction. The screen, colored a neutral gray, is easy to see through from the inside yet affords daytime privacy. It mounts like conventional screening. The cost is moderate.

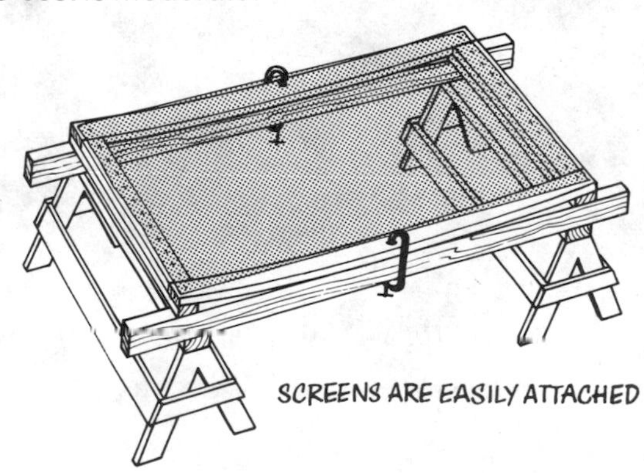

SCREENS ARE EASILY ATTACHED

Saran shade cloth for sunlight control

Saran is a plastic mesh manufactured in different densities of weave for precise control of light transmission. Although aimed primarily at the nursery trade, it has also been used successfully as a patio cover in certain mild climate regions, particularly in Southern California.

Because it is plastic, it will not rust or corrode. It melts at about 270° but will not support combustion. The greatest deterioration factor is abrasion.

Saran comes either uncolored and translucent or in dark green. Fading is negligible. You can buy weaves to give you from 30 percent to solid shade, as well as a lath-type weave with closely woven and comparatively open strips alternating. Saran is available in widths from 28 to 48 inches and in some weaves up to 20-foot widths and 900-foot lengths.

REDWOOD FRAMES hold saran shade fabric. Panels slide under each other, varying light, shade.

This particular plastic shrinks about 2 percent during the first two months of exposure to hot sunlight, so it should be installed loose. Although the ultimate life of saran is not known, its durability and weather-resistant qualities have long been familiar to outdoor furniture makers because it is commonly used for webbing in patio furniture.

Wire mesh and fencing

In this category are the many woven wire products that can be used to add a touch of texture and relieve the monotony of some large overheads. Some of the woven wires provide an excellent foundation for growing vines.

Hardware cloth, metal lath, chicken wire stretched taut, expanded aluminum and steel, and a variety of fencings give you a wide selection; but make sure the material is galvanized or otherwise treated against the weather before you buy.

Installation is simple if the mesh is stiff and precut to size. Just nail it down with common galvanized staples. With a flexible mesh, you must take care to stretch it tight.

You may have to hunt around a bit to find the mesh you want. Best places to look are building supply firms, metal specialists, fencing companies, and hardware stores.

Louvered screens are special orders

Two types of screens with tiny louvers built in are not usually available at stores where screening is sold. They must be ordered by the retailer. One type is the one-piece aluminum sheet with 17 louvers stamped into each running inch of sheet stock. The other is a sheet made of very narrow strips of steel or bronze woven together with fine wire, like a miniature venetian blind. The woven screen has 25 louvers to the running inch.

The bronze and steel cast a denser shade than the aluminum because less light is reflected off the louver surfaces. The aluminum is given a special treatment that helps resist weathering effects and gives it a pale, greenish cast; the steel and bronze are protected by a dark plastic coating.

Steel and bronze screening costs two to three times as much as aluminum, and according to professional screen men, it is trickier to install properly. Both materials can be purchased already installed in metal frames, but the cost is double or triple the cost of unmounted pieces.

Aluminum screening can be installed by the same methods as flyscreening; but the woven louvered screening can easily be misinstalled, so it would be wise to get a thorough briefing from the dealer before you try to install it yourself.

THE CHALLENGE here was to create an outdoor living area on a hillside. Existing concrete block retaining wall around an angular piece of land was used to build totally screened deck area. (Architect: Jon F. Myhre.)

Screens make you unbuggable

GARDEN AND POOL lie within screened enclosure (more than 1,100 square feet). (Landscape Architect: Phil Fields.)

VENTILATION comes from sliding glass paneled overhead. (Architect: Jack Hermann.)

Lath, batten and lumber

For sheltering plants or people, choices are limitless

DOMED GAZEBO is built with redwood lath and 2 by 2s.

Perhaps one of the oldest and certainly one of the most versatile overhead covers is wooden lath. Inexpensive, easy to install, and adaptable enough to provide you with as little or as much protection as any other cover, lath deserves consideration for any overhead where water-tightness is not a requirement.

Strictly speaking, laths are thin wooden strips which, when nailed to the framework of a building, are used as a surface for plastering. But long ago the outdoor builder borrowed a bundle of lath and tacked it to a frame to protect his shade-loving plants and himself from the heat of direct sunlight. The striped shadow pattern cut the withering effect of the sun's full blast but let in enough light and warmth to encourage some plants to thrive and to make the outdoors a pleasant refuge on a hot summer day.

Because of the unlimited number of ways that lath can be spaced, staggered, and patterned and because many sizes of wooden strips can be used, design may vary from a simple rustic trellis to the smartest contemporary patio shelter.

DESIGNING WITH LATH

Some of the most impressive lath overheads are made of one size lumber—1 by 1s, 1 by 2s, or 2 by 2s—with uniform spacing, but there is no reason why you cannot mix sizes and spacing. The only danger lies in overdoing it. Here are some examples:

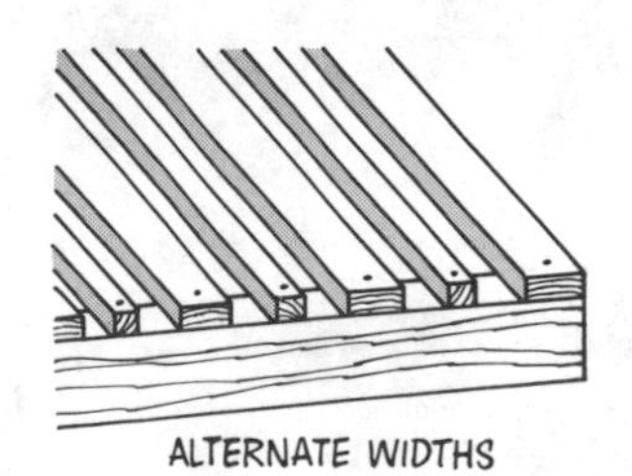

ALTERNATE WIDTHS

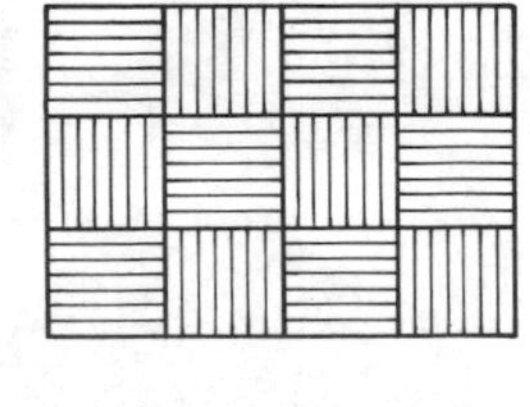

CHECKER BOARD

PURPOSELY DESIGNED for sun and shadows, this roof features redwood.

In addition, a number of shade patterns can be created by staggering the laths on both sides of the rafters.

Kinds of wood to use

By far the most satisfactory woods to use for outdoor lath construction are those that are naturally weather resistant, such as redwood and western red cedar heartwood. These woods have built-in resistance to decay, they do not require painting, and their straight grain makes them less liable to warp or twist under the merciless attack of the weather. Alternating heat and cold, dryness and wetness, or sun and shade tend to make other woods shrink, twist, or sag.

If you are unable to obtain either of these two woods, you might be able to get by with pine or fir, provided you are willing to go to some trouble to protect them against the weather. If you leave them unprotected, the laths are liable to warp and sag to a degree that will spoil the appearance of your overhead roof in a couple of seasons. To get around this tendency, you can apply a sealer-preservative, either by brush or by soaking in a trough.

Sealer-preservative may also benefit redwood and red cedar and is necessary if your batch of lath contains sapwood or if its light color indicates that it may be lacking in natural preservatives.

Ask your dealer for preservatives containing either pentachlorophenol or copper naphthenate plus a water sealer. Note that some preservatives are irritating to the skin and damaging to plant materials, so handle them with respect.

Sometimes you can buy small-dimensioned lumber that has been pressure-treated with preservative at the mill. Although this is somewhat more expensive than the untreated grades, it will provide you with long-lasting protection. Some brands of roll fencing that are usable for lath overheads are so treated at the factory; they will require no further protection.

Application of a stain will protect wood, add the desired color tone, and in many cases eliminate the need for the sealer-preservative treatment described above. Painting is not advised; not only is it a tedious chore to paint several hundred strips of wood but also you will be faced with the probability of having to repeat the job whenever the paint begins to crack and peel.

SIMPLE DESIGN of redwood posts, beams, and rafters shows use of post anchors, post-beam connections.

WESTERN CEDAR overhead extends fence design—wood roof for a shady area and plastic for sunlight.

Types and sizes

When you walk into the lumber yard to buy lumber for the lath cover, you have a number of choices. Here are the most common ones:

Lathhouse lath. Common outdoor lath is rough-surfaced redwood or cedar, milled to dimensions of about 3/8 by 1½ inches and sold in lengths of 4, 6, and 8 feet in bundles of 50. Two grades of lath are milled, but the lower grade has too many imperfections to make it a satisfactory covering material.

Batten. These overgrown laths are milled in thicknesses of ¼ to ¾ inch and widths of 2 to 3 inches. Unlike lath, batten can be purchased in lengths up to 20 feet and is generally sold by the piece rather than the bundle. Battens are sometimes cut to 6 or 8-foot lengths, however, and sold in a package of 30 or more.

Although smooth-surfaced batten is sometimes called lattice, it is not sold under this name in all localities.

In your search for thin boards, don't overlook the lumber yards that specialize in fencing materials. Many carry a good selection of small-dimensioned lumber. Although it's not obtainable everywhere, basketweave fencing, ½ by 4 inches, or ½ by 6 inches in 20 foot lengths, will make suitable covering, either as a woven ceiling or as parallel strips. If installed like lath, these boards are liable to warp unless supported about every 2 feet.

Boards and framing lumber. Heavier members can be used in your shelter, and although the cost may be greater than for lath, the structure will appear to be more solid and permanent—as in fact it is.

Generally, 1 by 4 and 2 by 2 rough lumber is easily obtainable at lumber yards, but frequently you can also find 1 by 2s, 1 by 3s, and boards of thicknesses 1⅛ or 1¼ inch. If these sizes are not in stock, for a slight extra charge the lumber dealer will rip stock 1 by 6s or 1 by 8s to make 1 by 2s or 1 by 3s.

Strips of 1 by 1 make a pleasing sun filter, but these usually have to be ripped from wider boards. You may be able to obtain 1 by 1 bean poles already cut, but don't count on it.

Surfaced lumber is sometimes used to add a polished quality to lath construction. Surfaced lumber costs more—from 5 to 10 percent—and the dimensions are smaller than the nominal size. The amount of surface removed varies with the size of the piece, but you can generally figure ½ inch off the width of the lumber you would use for lath. For example, a 1 by 4 surfaced on all four sides actually measures ¾ by 3½ inches.

When the lumber yard rips wider boards to furnish 1 by 1, 1 by 2, or 1 by 3-inch stock, remember that one or two sides of each piece will be rough unless the boards are planed off after cutting.

There is no reason why boards or framing lumber of greater width or thickness cannot be used, but 1 by 1, 2, 3, and 4, and 2 by 2s provide a selection wide enough to fill the requirements of most builders.

Grapestakes. These fence builders' favorites, which are roughly 2 by 2-inch split pickets, can be used to build a rustic but sturdy overhead. Grapestakes are usually available in 6-foot lengths and cost about 25 percent more than the equivalent rough 2 by 2s. Split grapestakes, which approximate 1 by 2s, make a satisfactory cover, but their varation in size and thickness produce an even more rustic appearance than the unsplit.

Wood and wire utility fencing. This material, which is sometimes known as snow fencing, is manufactured from surfaced fence pickets, woven together in 50 and 100-foot rolls. Standard widths are 25, and 18 inches.

Although the size of the lath varies, the wood is usually cut ½ by 1½ or 2 inches with 1 to 2-inch spaces between each piece. If the wood is not naturally resistant to the weather, it normally will have been treated with a preservative during manufacture. In addition to the protective treatment given it, some roll lath fencing is stained or painted and is obtainable in a limited number of colors—green, red, white, and redwood stain.

The greatest advantage of this material over common lath is the lower installation time. With lath fencing, the lath is pre-spaced and is considerably easier to attach. Instead of nailing down each piece, you can install the roll in complete sections. Usual practice is to rest both edges on cleats nailed to the rafters, as in the drawing.

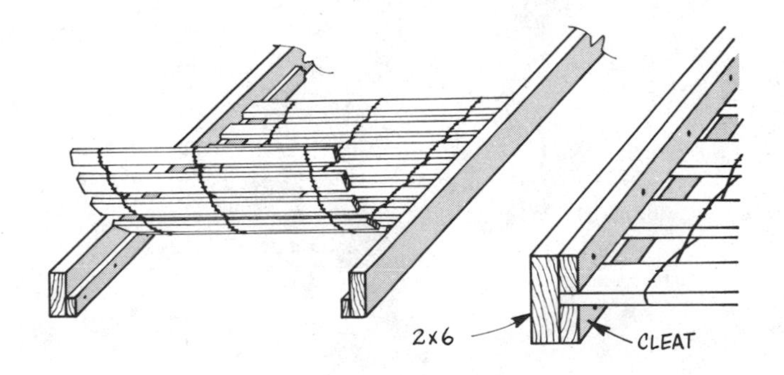

Some nurseries use this fencing overhead to protect their plants, but the spacing between the laths may be too wide to allow for comfortable living on the patio.

Blocking the sun

In order to plan for the shade your lath shelter will cast, you should take into consideration orientation, thickness, and spacing of the lath members. The time of year, the latitude in which you live, the exposure of your patio shelter, and the fencing or natural growth around the patio will also affect the amount of shade. But by carefully selecting the lath and controlling the direction in which it is installed, you can go a long way toward getting the effect you want.

Orientation. Ideally, you would have to be able to run your lath in any direction from north-south to east-west to have a perfect choice, but most builders will have to run the lath parallel or at right angles to the wall to which the shelter is being attached. To decide which direction the lath should run, make up your mind about the time of day you want the maximum shade from louver effect that the lath may cast. If you want the greatest relief from the sun at noon, run the lath east-west; if you want the greatest relief in the early morning and late afternoon, run the lath north-south. If the patio is so exposed that the lath can run only southwest-northeast or southeast-northwest, you will get the maximum relief in mid-morning or mid-afternoon, respectively.

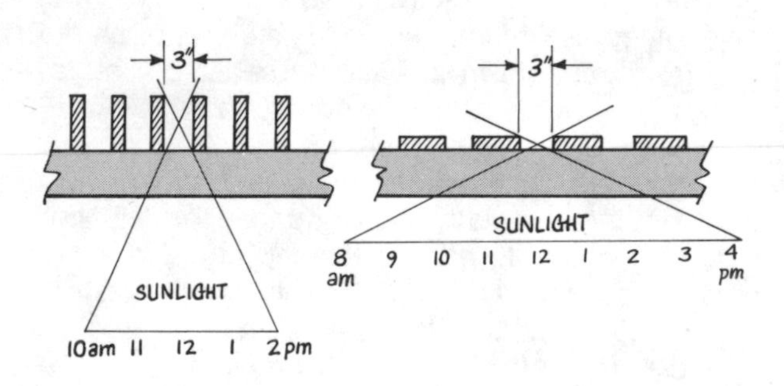

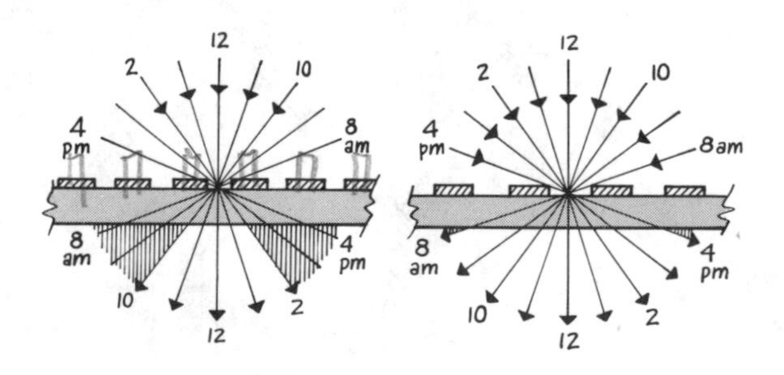

Since the louver effect will vary with the thickness of the lath and its spacing, you will have to decide on these two factors before you make the decision to run the lath one way or the other. If you plan to use 3/8-inch-thick lath or batten spaced an inch apart, you can expect a minimum louver effect. However, if you plan to use 2 by 2s spaced half an inch apart, the louver effect will be considerable, and it would be well worth the time to figure the most advantageous arrangement.

Spacing. For a pleasant shadow pattern regardless of the time of day, you can follow the practice of landscape architects, spacing the lath according to arrangements they have found successful. For lath 1/2 inch in thickness or less, the spacing should be from 3/8 to 3/4 inch. For lath from 1/2 to 1 1/8 inch thick, the spacing should be from 3/4 to 1 inch. For 2 by 2s the spacing could be as wide as 1 1/2 to 2 inches under some circumstances, but 1 to 1 1/2 inch will make the patio more comfortble in most cases.

Remember that the higher the patio roof, the more diffused the light becomes. The closer the roof, the sharper the striped shadows.

The amount of space between each member should not be governed solely by the width of the lath. Where 1-inch spacing with 1 by 1s may produce a delightful effect, 1 by 4s with 4-inch spacing would produce an irritating impression.

If you are uncertain about the best spacing or orientation of the laths of your overhead, you can easily experiment for a day or so and find out for yourself what is the most pleasing combination. After you build your supporting frame, tack on two or three panels of lath spaced different distances apart. Drive nails into the lath just far enough so slats will not blow away—but not so far that you cannot get the hammer claw under the nail heads. Try out the lath spacings for 24 hours or more to see which orientation and spacing will work out the best. It may take several trial runs before you get the most pleasing combination, but this is preferable to nailing down a thousand lath and later finding that the zebra shade they cast is unpleasant or that the shade you want comes when you don't need it.

MATERIAL REQUIREMENTS

Once you have figured the total area of your patio roof, the size of lath members, and the spacing, you can determine the amount of material you'll need. The material you buy will be sold in bundles, by the running foot or by the board foot. Lath is usually sold in bundles of 50; battens, 1 by 1s, 1 by 2s, and 1 by 3s by the running foot; and larger sizes by the board foot (1 by 12 by 12 inches = 1 board foot). Broken bundles or individual pieces are more costly. Computing your needs in board feet is complicated and unnecessary; if you provide your lumber dealer with lineal feet requirements, he will quickly convert it to board feet.

To figure how much lath to order, you will need to follow a simple series of computations:

1. First, find out how many running feet of lath are required per square foot of patio roof. To determine this add the width of the lath and the space you plan to leave open between laths and divide the total into 12. For example: for 1½ inch lath spaced ½ inch apart, add 1½ and ½ and divide into 12, giving the answer six.

2. To find out how many running feet are needed to cover the area of your patio roof, multiply the running-feet-per-square-foot times the total number of square feet in the overhead. For example: if there are 6 running feet of lath per square foot and 100 square feet in the roof, you would need to order 600 running feet of lath.

If lath are sold by the bundle in your locality, you can determine the number of bundles to order by simply figuring the number of running feet per bundle and dividing this into the number of running feet required. The number of running feet per bundle is easily determined by multiplying the length of the lath times the number of laths in the bundle. A 50-lath bundle of 4-foot lengths, for example, would contain 200 running feet. Three bundles would cover a 600-foot roof. On the other hand, a 50-lath bundle of 6-foot lengths would contain 300 running feet and only two would be required for the same roof area.

Tools

In all probability you'll need nothing more than a hammer and a saw to install the lath. Be sure to use hot-dipped galvanized or aluminum alloy nails to secure the lath to the frame. With ⅜ or ½-inch-thick lath, use three or four-penny nails, either box or common. Eight-penny box nails will do the job with 1-inch stock, as will 12 or 16-penny with 2-inch boards.

Framework

Most builders who put up a lath overhead want a permanent installation. The structure recommended in the construction chapter (see page 18) for 30 pounds per square foot load is adequate to carry any lath cover with no danger of sagging or structural failure.

Spacing. The problem of getting uniform, parallel spacing is easily solved. After you have nailed down the first lath, lay a board exactly as wide as the space you want, push the next lath up against the guide board, and nail down both ends of the lath. It's a good idea to use two nails on each end of the lath. A tendency to warp or curl will be checked somewhat by the secure attachment.

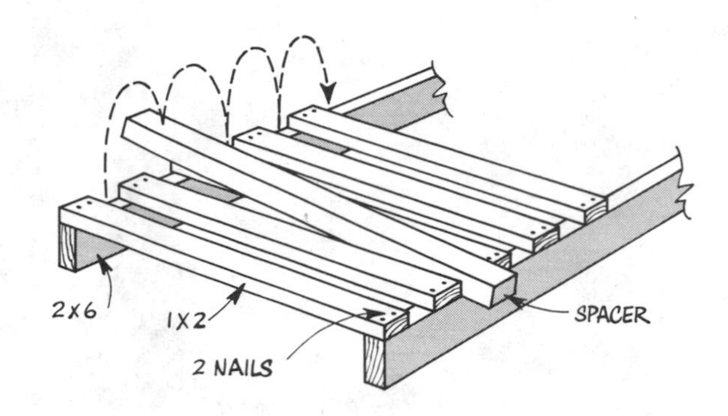

Spans. If you want to avoid putting up a lath cover that will sag or warp, it's best to be conservative on the distance you span with the lath.

For common lath and batten, 2 feet is the maximum; with 1-inch stock you can span up to 3 feet but 2 feet is better; 1 by 2-inch stock on end and 2 by 2s will span 4 feet without objectionable sagging. A certain amount of irregularity in the lath cover is pleasing, and you can count on the boards to provide it for you as they weather; but be careful to nail the lath down evenly spaced and in perfect alignment. Misalignment, sagging, and uneven spacing are unattractive, but a little bit of twisting or bending relieve the severity of a geometrically perfect installation.

Frames. If you decide to use removable frames instead of making a permanent overhead, there are several ways of doing it. Naturally, there are an unlimited number of sizes that could be built, but we suggest 3 by 6 feet as being close to the optimum. That's about as big a panel as you can handle easily, and by making the panels large, you save on materials and time.

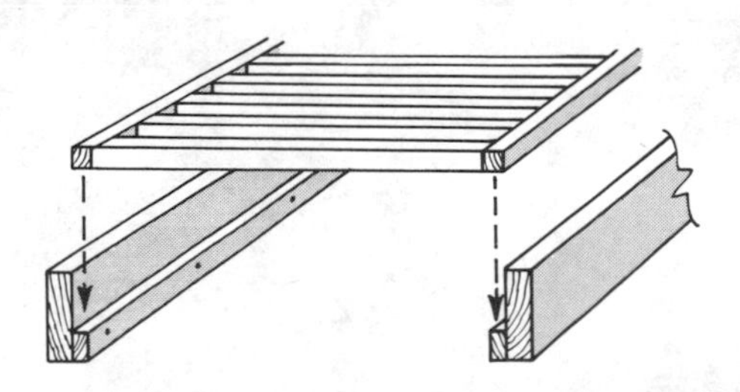

Five ways to design a roof

BIG DECK for hobbies and entertaining built from 4 by 4 redwood posts and 2 by 6 rafters mounted on 4 by 12 beams to support roof. Deck is garage roof with storage cabinets and sculpture bench.

LEFT: simple lath and batten structures of the type shown here have excellent records as garden shelters. BELOW: decorative use of lath over deck.

GENEROUS USE of finished lath in overhead enhances open feeling of this spacious poolside dining, recreation, and garden area completely independent from house.

FINISHED LATH is supported by 2 by 6 rafters, 4 by 4 posts and beams; has 1 by 8 border.

The canvas cover

You can sew it, lace it, adjust it, or store it

POOLSIDE CANOPY has gaslights attached to metal framework.

Canvas shares one similarity with lath: each had humbler uses before the outdoor builder started looking for materials to block undesired heat and light from his patio or terrace. There, the similarity ends.

At one time the only material used for awnings was canvas. It is still judged the best fabric for most outdoor installations. The traditional pull-up, striped awning that is used to shield house windows from the sun is today supplemented by sheets of solid-colored canvas laced to pipe or lumber frames or strung from cables for adjustable weather control.

KNOW YOUR CANVAS BEFORE YOU BUY

Canvas comes in a wide variety of weights, finishes, weaves, and colors, but because sun, wind, rain, and mildew may under some circumstances hurry a stout piece of canvas to an early grave, careful shopping is advisable.

Although a variety of weights and weaves are suitable for outdoor uses, awning experts recommend a 10.10 ounce fungus-proofed Army duck as the best for use on an overhead. (See *Mildew protection,* next page). This chemically treated Army duck comes in rolls of 31 inches and in rainbow-hued solid colors and stripes.

Canvas weights vary from 7 to 15 ounces per yard. The consensus seems to be that lighter canvas cannot take the physical abuse of outdoor use, and the heavier canvas requires longer to dry out after it has become wet and thereby is subject to mildew. Because of its chemical treatment, Army duck is a special weave and weight designed for outdoor use.

There are three ways canvas is colored: painted, vat-dyed, and yarn-dyed. You can also buy plain, off-white and pearl gray (colorless) canvas everywhere the material is sold.

SHADE of color bordered canvas strips casts delicate pattern on granite floor.

Painted canvas, available in a wide range of colors and patterns (stripes, plaids, checks), is preferable for outdoor use. The acrylic paints used today are far superior to the old oil-based paints in providing water repellency, weather resistance, and color fastness. But make sure that the canvas has been treated to prevent fungus growth.

For durability, the vat-dyed canvases are least suitable for outside use. They fade more quickly than the yarn-dyed or painted materials. The best dyed canvas may, in certain colors, resist discoloring or fading better than some painted colors, but this depends on the quality of coloring agent. A new acrylic fiber is guaranteed against fading (except red and yellow) for five years. Care should be taken in application of these fabrics, however, because they stretch when wet and water collects in puddles on the material. Large spans or flat installations are not recommended. This material does make attractive pads, roll shades, curtains and umbrellas, but the price is high.

Some find the pearly gray or flower-printed underside of the painted canvas unattractive, but that surface is in the shade and not very noticeable in most overheads.

The neutral, off-white canvas is popular because it doesn't conflict with a color scheme or detract from a colorful garden, and it blends with natural surroundings better than the dyed or painted canvas. One canvas expert remarked that sitting under the off-white material is like sitting beneath a billowing sail. The light that shines through is white and non-glaring.

Another variety of canvas is the vinyl-coated material which, although it costs slightly more than other types, is longer lasting. It is a particularly good choice for whites and pastels because it is sun-fast and will shed dirt better than the others.

How long will it last?

Generally speaking, the life of a fungus-proofed, painted canvas overhead given good care should be a minimum of five years. Average life is from eight to twelve years. There is, of course, no way to predict exactly, because of the many variables that affect canvas life. Under adverse conditions, untreated canvas may not survive more than one or two seasons. Humidity, sunlight, exposure, and lesser factors affect the life of the canvas. For instance, an overhead located near or under a tree will mildew more quickly than one that is not. Leaves will lie on the top surface and act as tiny sponges to provide a bed for fungus growth.

The relationship of the canvas to the sun will radically affect the life of the fabric. If the cover is in direct sunlight and free of damp leaves, it will far outlast a piece installed on the north side of a building where it receives little or no sun. The reason is simple: the sun dries dew or rain from the canvas quickly and prevents or greatly retards the growth of mildew.

Mildew protection

If you expect the canvas you put up outside to give you many years of service, make sure it is fungus-proofed. The life of the canvas is greatly extended in outdoor use when thus protected. The chemical treatment applied to canvas for outdoor use is an effective anti-fungus agent. It works in two ways. The treatment blocks attack by exterior fungus growth and it prevents development of interior fungus growth during seasons when the canvas (which acts as a sponge) remains wet or damp for long periods. Ask your awning dealer about the various patented treatments, or write to the Canvas Products Association International, 600 Endicott Building, St. Paul, Minnesota 55101.

Maintenance

Occasional hosing accompanied by a light brushing to remove dirt, leaves, and twigs will help to prevent mildew. If your overhead is retractable, leave it down when it rains—the cleaning will do it good. Whenever you do retract the canvas, make sure there is no water, dirt, or other residue in the folds.

Don't let vines or other plants come into contact with canvas (moisture from them can cause mildew).

Rips or punctures can be repaired with fabric cement and a patch of canvas. A top-surface patch looks best, but you'll get a stronger patch if you apply it to the underside. Leave a minimum overlap of an inch; for larger splits, allow 3 inches.

When storing canvas, first clean it and let it dry thoroughly. Never store it on concrete or earth floors where it can absorb moisture.

Other fabrics

Various synthetic fabrics have been introduced during the past few years, but it is too early to tell whether they will eventually be able to match the all-around performance of canvas as canopy and awning material.

Neoprene or vinyl-coated nylons have been used successfully for pool covers and truck tops. Be sure to ventilate, however, or you will have moisture condensation problems. Saran cloth (see page 36) is an effective overhead for certain situations.

Cotton sheeting, denim, and burlap are light in weight and inexpensive, but their life span is much shorter than that of canvas.

SEWING AND GROMMETING

Recently one of the old complaints against canvas awnings—that the thread used for sewing the pieces together and for stitching the hems rotted away before the canvas life was spent—was remedied by the discovery that Dacron thread had all the qualities needed to outlast the canvas itself.

You may wonder why Orlon fabric is not used for covering overheads. It has been tried, but experimental installations have shown that because it won't shrink, the cloth is difficult to fit snugly in place and it looks permanently wrinkled. Unlike canvas, Orlon will not close up and become watertight in a downpour; instead it acts as a sieve.

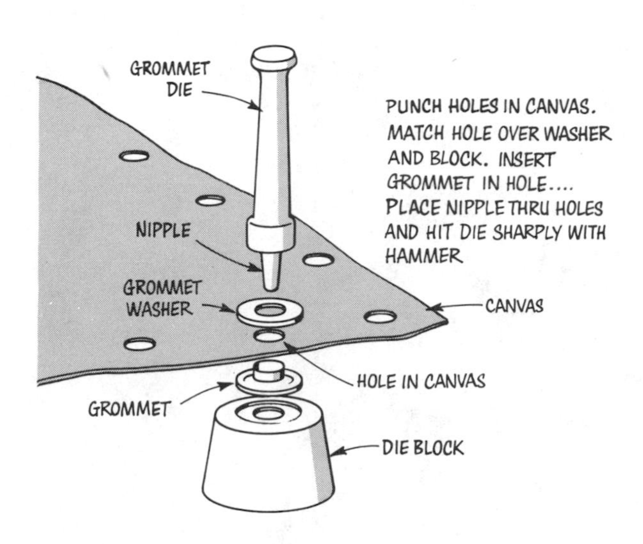

Although you can sew 10-ounce duck on your sewing machine, if it is vinyl-coated or heavily painted, you had better entrust its hemming to the awning shop. Canvas heavier than 10-ounce weight must be sewn with special equipment.

To hand sew canvas, use a Number 13 sailmaker's needle. If you have the grommets installed by the awning shop, they cost about ten cents apiece, or you can do it yourself, as shown, at some savings. For a lace-on cover, the grommets should be placed every 8 inches. Make sure the grommets are so placed that you wind up with one in each corner.

Use a rope or cord the diameter of ¼ inch (venetian blind) cord to lash the canvas to the frame. If you don't like the look of laced-on canvas, you can retain all the advantages of that type and still have your cover look like the slip-on sleeved canvas. Just have the canvas cut so that you can lap the edges around the pipe frame and lace them together across the top.

Lace-on canvas covers

The trim nautical look of canvas pulled taut in pipe panels has captured the imagination of many, and although there are other quite satisfactory ways of installing it, the lace-on method has other advantages that may help account for its popularity for installations.

First, installation is quite simple. Wrestling with a bale of canvas and half a dozen pipe members, as you will if sleeves are sewn into the cloth, is a good deal more work than roping the panel to a pipe frame.

Because tension on the canvas tends to be evenly distributed when it is laced on with a continuous piece of cord, wrinkling and bunching, which may occur when an inexperienced person installs the slip-on type of cover, can be avoided.

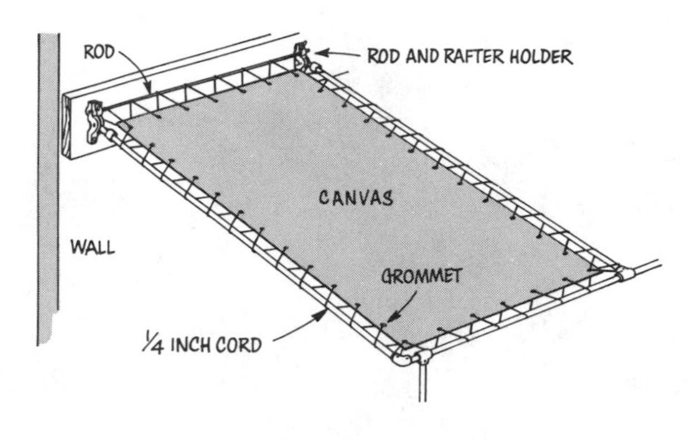

Finally, maintenance is simpler. The canvas can be removed for the winter months with little trouble; tension can be adjusted to keep the panel tight and wrinkle free; and if you don't mind the sight of it, you'll never have to dismantle the frame.

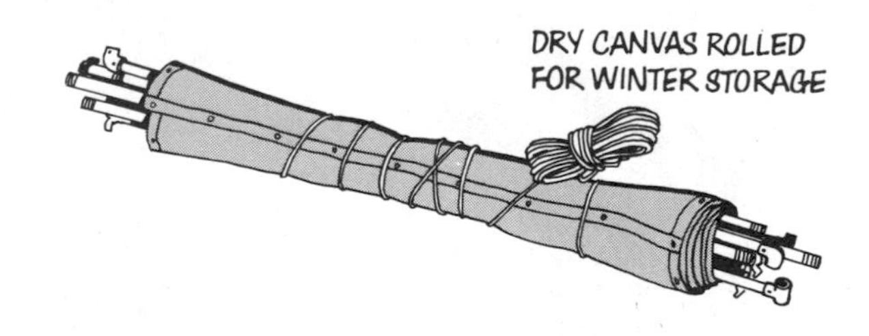

Adjustable canvas cover

Another method of supporting a canvas overhead is to suspend the fabric on strong cables. With this system, the canvas can be shifted around over the patio area. The suspended covers are usually made up of a series of 5-foot strips, running parallel to each other. Wider pieces can be used, but if they are wider than 10 or 15 feet, they may be cumbersome to move back and forth on the cable and are liable to sail off in a strong breeze.

Your awning man may not have had much experience with this type of installation, and in that case, a proper cable can be selected by a hardware dealer or smallboat rigger familiar with the materials. The cable can be attached to the house or frame with pad eyes (large screw eyes) or awning hinges. It's a good idea to use a turnbuckle at one end of the cable to take up slack from stretching.

The canvas should be suspended by rings attached to it either by baby snaps sewn to the fabric or by cotter pins passed through grommets and spread out underneath.

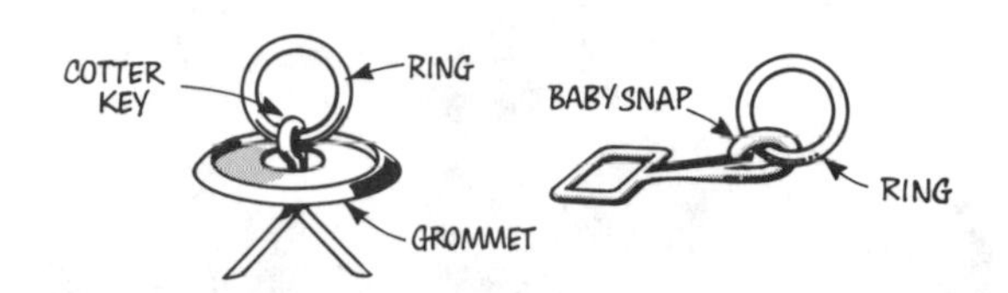

Basket-weave cover

Another way to use canvas as a cover is to weave strips, leaving open spaces (see photo on page 46) for some light transmission and the free circulation of air.

Have the material cut and hemmed at an awning shop to the length needed. The standard width for most canvas is 31 inches, so it would be wise to consult with the dealer on the widths of hemmed strips he can make from the roll without waste. Each strip will require two grommets at each end for tight, stretched installation. The strips can be hemmed, or they can be bound in a different colored cotton tape to add interest.

THE PIPE FRAMEWORK

The only complicated thing about building a canvas overhead is erecting the pipe structure and making it secure. Fortunately, all the parts are commonly available and simple to use. New slip fittings and simplified frameworks that can be used in place of the traditional threaded pipe frame parts make it easier for the home owner to build the structure; but though they are easier to work with, they are not so sturdy as stock threaded parts and may not be adequate for some installations.

The pipe framework is fitted together with standard parts that are easy to obtain. Some of them come from the awning shop, some from the corner hardware store.

Posts, rods, and rafters are standard galvanized water pipe, usually ¾-inch, that can be bought in any length at almost any hardware store. If you do not have a threading machine, you can have the pipe threaded at the store after it is cut.

Special fittings to join parts of the framework together and attach it to the house are obtainable from your awning shop. These parts, as shown in the drawing, are 1) rod and rafter holder for attaching to the house; 2) eye ends for attaching pipe to the rafter holders; 3) tapped els for right-angle corners; 4) slip tees for joining three pieces of pipe in a T-joint; and 5) post plates or flanges for supporting the base of the post.

You would be wise to let your awning shop figure out the pipe frame, but you can get a rough idea of the general requirements from the drawing below.

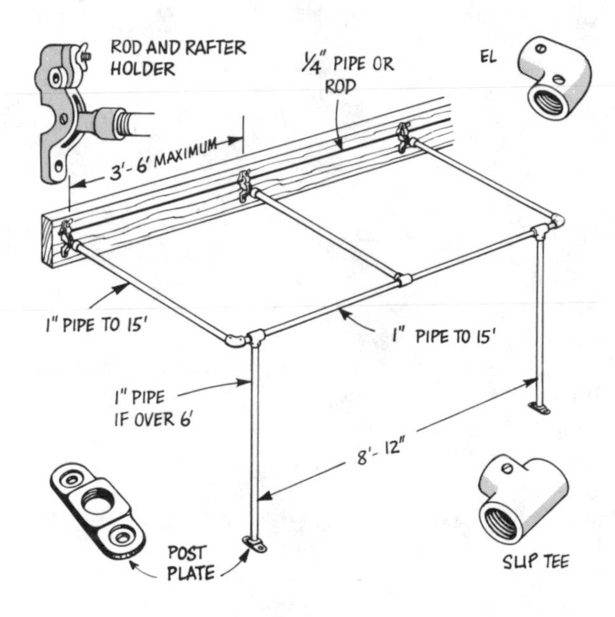

Regardless of the type of canvas cover you put up, it is advisable to limit the size to about 10 by 10 feet for each piece. A 10 by 20-foot overhead is less likely to be carried off in a big wind if it is made of two 10 by 10 or four 5 by 10-foot sections than if it is made in one piece. The open spaces between laced-on canvas and the frame help resist the tendency of the large flat area to act as a sail.

Don't plan to cover an area of more than 5 by 10 feet with one panel if you expect to avoid sagging and flapping entirely.

RETRACTABLE CANVAS panels can be folded back in late afternoon when tall eucalyptus shades deck.

Three ways to canvas

WEST SIDE CANOPY shades floor to ceiling windows of house at left, out of picture.

TAILOR-MADE and laced to pipe frame jutting from roof overhang, canvas doesn't project weighty feeling that a heavy roof does. Treated with preservative, it takes year-round weather for five years or longer.

Glass, plastic and aluminum

Some install easily, others need professional skill

SWIMMING is all year. Sliding plastic panels allow ventilation.

All three materials (glass, plastic, and aluminum) can be incorporated into the same roof to allow for simultaneous sunlight, partial shade and total shade. Each can be encased in movable (for ventilation) or removable (for storage) covers.

If you design your patio roof according to the standard sizes sold by most dealers, construction will usually go along much more easily. Accessories are manufactured for standard panels.

Both corrugated or crimped aluminum and plastic panels are attached to the rafters in the same manner. These materials can in fact be ordered in identical lengths and installed alternately, overlapping one another. Both aluminum and plastic panels come in an attractive range of colors and shapes.

Here, under individual headings, are the characteristics of each material.

THE GLASS OVERHEAD

For many decades, glass has been used to bring the out-of-doors inside. If a builder wants warmth and an unobstructed view throughout a rainy summer—and the added benefit of permanence—he can do no better than to build his room with glass walls and ceiling.

If you have a patio with a north exposure and you want all the light you can get, yet want to keep the rain away, glass may be the thing for you.

For some patio shelters, glass has no equal, but if it is misused, the glass roof can act as a heat trap and pose a condensation and drip problem, making life beneath it most uncomfortable.

Unlike polyester plastic (the only other widely used translucent material that affords positive protection from the elements), glass is difficult for an

WHITE PLASTIC ROOF PANELS, clear glass doors enclose garden-room.

unskilled amateur to handle and install with assurance of watertight permanence.

There is no sense in putting up a glass roof if it is not done right, and horizontal glazing is just not one of those things that most week-end builders can master the first time out.

Two kinds of glass roof construction can be used that fill the basic requirements of permanence and weather-tightness: conventional skylight and the familiar greenhouse methods.

Skylight construction

If a roof is built to specifications for skylight construction and the area is ventilated or otherwise protected against the heat and condensation problems, it should give a lifetime of service. The cost is relatively high for this type of construction, and the work should be done by someone who knows the materials and is familiar with the problems that can arise when you enclose a chunk of atmosphere in glass.

Normally, you would use wire-reinforced, 1/4-inch glass, but there are many other types that can be substituted. Textured, obscure glass, heat-resistant glass, and tinted or patterned panes can be used to meet your special requirements.

Double-pane construction and special aluminum sash can help meet the heat and drip problems.

The place to get the information you'll need for planning is a store that specializes in glass. The people there will know about the materials that are available and construction requirements, and they will be able to recommend a general design that should forestall trouble with heat and condensation. Of course, an architect or an engineer can also provide you with the proper information.

Greenhouse roof construction

Although a steeper pitch is needed for a greenhouse-type roof, the construction is simpler and within the ability of some builders to do for themselves.

Greenhouse manufacturers usually use a rise of 6 inches for every foot of roof in order that condensed moisture will run down the panes to the little troughs in the special greenhouse glazing bar, which carry off the water that condenses on the inside surface of the glass. Condensation of moisture on a greenhouse roof is a natural phenomenon because of the moisture-laden air in the greenhouse.

A glass roof over a patio, however, is less liable to drip—unless the patio is fully enclosed. If you provide ample ventilation, you will not be bothered by a drip problem, and you can lay the glass on a more gradual slope than the steeply pitched greenhouse. If one wall of the patio is open to all outdoors, moisture is not liable to collect on the undersurface of the roof. What little that does, will quickly evaporate. But if the patio area is fully enclosed, you will need to provide good ventilation, for the drip can spoil the livability of your patio. You can ventilate the area with hinged roof panels, with shielded vents under the eaves, or with louvered glass windows installed in the walled area.

To make sure that you will not be troubled with this problem, consult an engineer, architect, or building inspector before installing the roof.

The typical framework for a greenhouse roof is formed of 2 by 3 or 2 by 4-inch rafters, double-rabbeted to hold the glass and to carry off the drip, as shown in the drawing below.

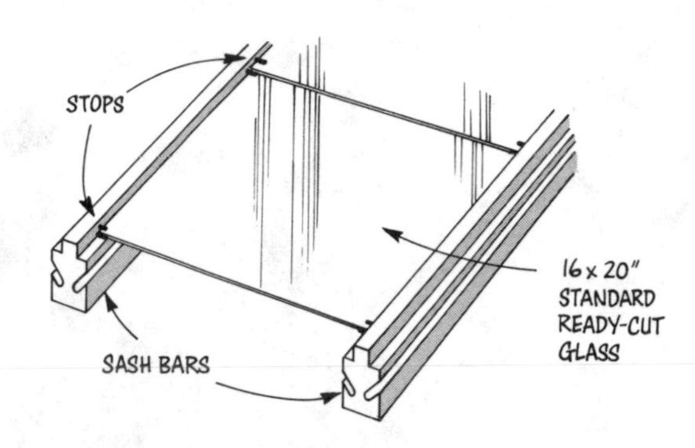

Rafters can be cut on a bench saw, or your lumber dealer may cut them for you if he does not carry them in stock. If you are sure of your ventilation and are pitching the roof to a gentle slope, you can do without the drain groove. For that matter, you can make up the rafters from 2 by 4s with 1 by 1s nailed to the top surface, leaving a 1/2-inch shoulder on either side, as shown in the drawing.

To save lumber and work, use the largest size glass you can work with. The ready-cut 16 by 20 inches works out handily. Let the glass size determine the exact measurements of the roof. For example, a patio roof that is approximately 10 by 12 feet will be 9 panes (16" x 20") long and 5½ panes wide. For the exact dimensions of the completed roof, add in the space between the panes. Thus

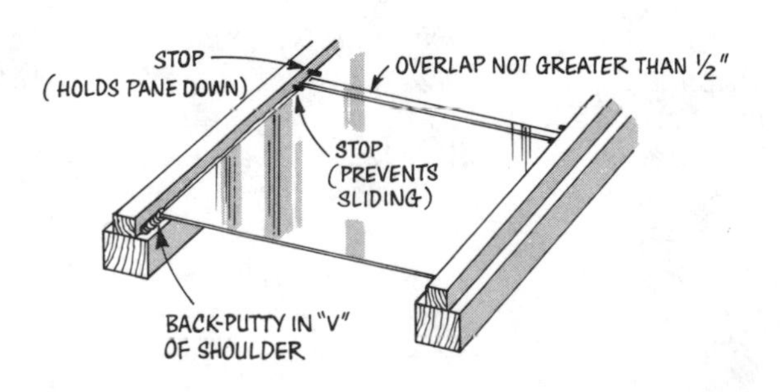

GLASS—BEAUTIFUL AND HAZARDOUS

ORDINARY GLASS shatters into sharp, dagger-like pieces when struck, can produce serious cuts.

TEMPERED GLASS breaks into sizable round chunks; is available in plate or sheet, clear or patterned.

The U. S. Public Health Service estimates that more than 100,000 Americans are injured, many of them fatally, each year by walking or falling through glass doors and panels. More than a third of the victims are children.

Most of these accidents can be prevented with the use of approved safety glass or plastic in hazardous areas.

If you have existing installations of untempered glass, you can still take steps to make sure that no one mistakes a panel of glass for an opening. (Even if you have safety glass, it is a good idea to take precautionary measures.) One way to point out that your glass door or glass screen is no passageway is to place furniture or plants in such a way that they control traffic. Another way is to place highly visible decals on the glass panels, or hang decorative mobiles or plants in front of them. Along a glass-screened entryway you might even install a handsome wood rail.

If you live in a community that has adopted adequate glass regulations—or if FHA financing is involved—the new house you buy, build, or remodel is required to have safety glass or approved rigid plastic in all hazardous areas. The materials that are permitted include fully tempered glass, wired glass, laminated glass, and rigid plastic. These materials will not break readily—and if they do, the chances of bad cuts are minimal.

1-inch spacers would add 10 inches to the over-all length of a 12-foot long patio roof covered with 16-inch wide panes.

Setting the panes. If you live in a region where snow may fall and build up on the shelter roof or where hailstorms are known, use double-strength window glass. Greenhouse makers normally use a B grade glass, but you may want to spend a little more for panes with fewer imperfections.

The bare wood should be primed with a good lead and linseed oil paint. When the paint is dry, putty or caulking compound should be laid along the "V" of the shoulder, exactly as back-putty is laid in window glazing. Beginning at the bottom of the roof, the glass panes are laid in the putty bed. Overlap should be not more than ½ inch nor less than ⅛ inch. (The water that is drawn up into the lap by capillary action may be blown in under the roof by gusts of wind. The bigger the lap, the more water will be blown—so be sure to keep it at a minimum.)

Fasten each pane in place with brads or finishing nails driven into the wooden strip on each side. One brad should hold the pane down, and another should prevent it from sliding downhill.

Free space of 1/16 inch must be allowed on each side between wood and glass. The plastic putty or glazing compound will squeeze up in a roll along the juncture of wood and glass as the glass is imbedded. This roll should be cut away flush with the surface of the glass by drawing the putty knife down the edge. Otherwise, cracks will occur in the roll and hasten deterioration of the putty.

The putty that lies under the glass will eventually dry and crack unless you use a mastic or caulking compound that won't set. If you use putty, a painting of the joints every three years will probably take care of any leaks.

KEEPING COOL

As with plastic panels, ventilation is a must if the glass roof is unprotected from the sun. You may try your best and still not keep the sheltered area as cool as you'd like it through the day. You will be better equipped to solve the problem if you understand a little about heat transfer.

Radiant solar heat is transmitted through glass just as light is. There's some reflection, and with regular window glass a very small amount of absorption, but the great part of the heat "shines" right through. Once the heat is inside, it is soaked up by the walls, floors, furniture, and other objects within the room which, in turn, heat the air. If the air is not free to circulate with the outside air, high temperatures will build up because the glass acts as a one-way filter. It will let *radiant* heat in, but it won't let the heated air out.

In addition to ventilation, there is another way to help keep the area under a translucent roof cool: keep some of the radiant heat from reaching the glass. If you keep all of the radiant heat out, you'll be blocking all the light, too. This can be done by painting the glass, usually white, or by hanging a shade or cover over the glass when you want to block the heat. The trick is to get the shade on the *outside,* not on the inside. An inside shade may help some, but the culprit is already in and you may find it too uncomfortable even with the shade.

Woven bamboo shades, canvas, muslin, and louvered screening are among the materials you can use to keep heat away from the glass.

STRONG WIND is reduced by panels of clear glass and tempered hardboard around two sides. To allow for cooling air circulation, open space is left both below and above the panels. (Landscape Architect: N. Bob Murakami.)

PLASTIC PANELS

In the relatively few years since they were introduced, plastic panels have developed into one of the most popular of all outdoor overheads.

The bright, colorful panels have many appealing features. Like other solid overheads, they will protect a patio from rain. The translucent types have the happy quality of letting through softly diffused light while cutting out the sun's glare.

Plastic panels are light in weight and easy to install, but the do-it-yourselfer should not interpret this to mean that he can tackle the job haphazardly. The rules for installation are indeed simple, but they must be followed carefully if good results are to be expected—even if you are using the very best panels obtainable.

How to buy plastic panels

The plastics industry has established minimum quality standards, but the consumer is likely to find these specifications difficult to interpret or to relate from one brand to another.

To make sure that you're buying the best material for the purpose, consider the reliability of the manufacturer and rely on the advice of a dealer whose integrity is known to you. If you purchase high-quality panels made by dealer-recommended manufacturers, you can expect many years of satisfactory use—far beyond the actual guarantee, in many cases. Inferior grades may show surface erosion in two or three years.

Most manufacturers provide free promotional material describing their products. It is wise to read it carefully, with attention to special factors as well as to the obvious ones of color, shape, and size. If you live in a cold climate, you should know whether the product will support a heavy snow load. Residents in fire-danger areas should give special attention to fire-retardant panels. Proven resistance to deterioration and weathering are two other important considerations.

Price is also an index, and, as with other products, the old maxim, "You get what you pay for," applies.

As a result of research and improved manufacturing techniques, the plastic panels of today are far superior to the first introductions of the late 1940s and early 1950s. Now, great diversification exists in the kinds and grades of plastic suitable for outdoor use.

Variety in plastics

Polyester resin and vinyl are widely used materials. Top-quality panels made of these plastics are reinforced with fiberglass for added strength. Some polyester panels include fiberglass matting, and a coating of acrylic resin (chemically bonded) for long life. Adequate coating with an acrylic will protect panels for up to 10 years.

Translucent (light-transmissive) types are the most popular. You may choose from several patterns: corrugated, flat, crimped, staggered shiplap, and simulated board-and-batten. Some are sold in both smooth and pebble-grained textures.

For those who want a light-proof overhead, smooth-finish opaque vinyl panels are now available in corrugated form. They also can be alternated with translucent panels for an interesting pattern effect.

(Polyethylene plastic, available in sheets from .0015 to .020-inch thick, makes a very inexpensive cover material, but only lasts a season or two.)

Sizes. You'll find many standard panel sizes, ranging from 24 to 50½ inches in width and 8 to 20 feet in length. Thickness varies by color and type. Plastic is also available in rolls up to 50 feet long; corrugated rolls are 40 inches wide; flat rolls are 36 inches wide.

If you design your patio roof according to the standard sizes sold by most dealers, construction will usually go along much more easily. Accessories are manufactured for standard panels.

Colors. Scanning the lists of panel colors offered by the various manufacturers can be a decorator's delight. Manufacturers' color descriptions such as "dusty peach," "mint," "tangerine," and "sandstone" are indicative of the wide range of hues.

Perhaps the most important thing to remember when selecting a color is that—unlike other colored overhead materials—translucent plastic panels are not only colorful in themselves but also will *cast* their color onto the patio and perhaps through a window wall and into the house. Therefore, you must keep in mind your overall color scheme, including the color of the house itself.

If you like a certain color but are doubtful as to how it will look, ask your dealer if he knows where you might see similar panels installed. Call the owner for permission to take a look, and go see for yourself. If you have the understructure up, perhaps the dealer would let you have the panels on a trial basis, and you could test them by laying them in place on the frame.

The different colors will transmit varying amounts of heat and light. Blues, greens, and some yellows are a good median choice for most situations. If you live in a hot climate, give special consideration to reflecting colors which transmit minimum heat and light; on the other hand, a clear or frosted panel may be the best answer for cool regions. Climate is not the only consideration—the orientation of the patio also plays a part.

How to install corrugated panels

Of the various patterns available, corrugated plastic panels are the most widely used for general outdoor overhead construction. Except for the overlap problem, most of the principles described below hold true for other patterns.

The most popular corrugated panel is 26 inches wide. This provides for a 2-inch overlap and installation on rafters spaced 2 feet on centers. The panels should be supported along the seams not only for strength but also for the sake of appearance. If the overlapping section is exposed, the less translucent seam breaks the expanse of diffused light without giving the sharp delineation that the rafters provide.

In addition, cross bracing is needed every 5 feet between the rafters to support the panels across the corrugations; without it, the panels may sag in time. The ideal installation for the 26-inch panel, then, would be on an eggcrate frame 2 feet wide and 5 feet long with the corrugations running lengthwise. For the many other sizes and patterns, the manufacturers have developed specifications (sometimes including special moldings) needed to make a sag-free installation.

Because of the wind loads that can build up, the supporting structure must be built to the specifications on page 19 or the equivalent (30 pounds per square foot minimum)—if you don't want to find your overhead over someone else's head after the first big wind.

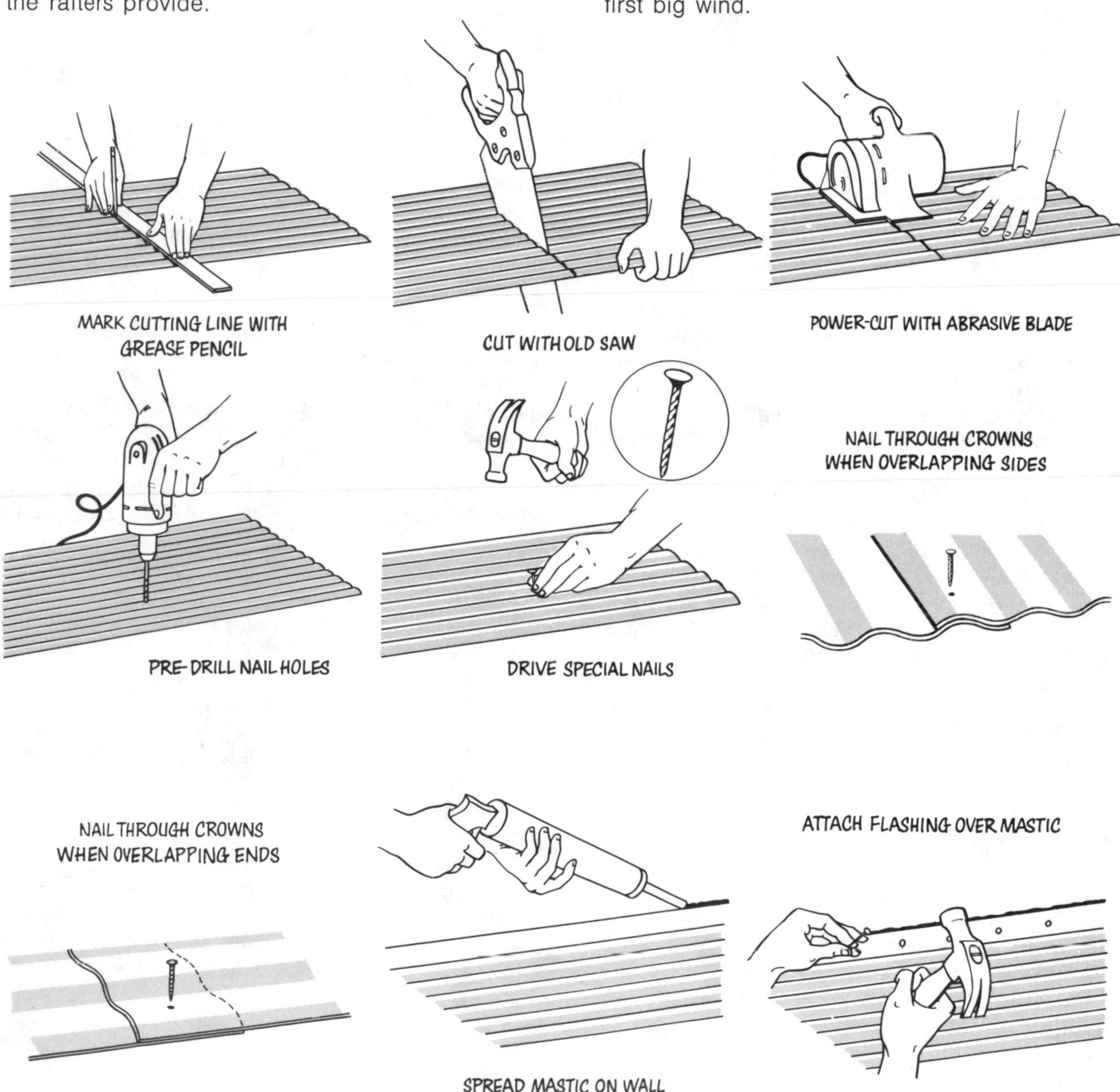

Although you can cut plastic panels with a saw, it would be well worth your time to build the overhead to standard dimensions and save the time and work required to saw the pieces. Where cutting is necessary, use a fine-toothed handsaw (preferably an old one) or a power saw with an abrasive blade.

Drilling and nailing are easily accomplished with ordinary power and hand tools. Some panels have a tendency to craze around the nail holes if the sheet hasn't been drilled previous to nailing, but the sheet won't be harmed. In most professional installations, pilot holes (slightly smaller than the nail diameter) are made with a high-speed drill to make nailing easier and give the finished job a neater look. Make sure the drilling surface is well backed to prevent your panel from damage.

Special aluminum twist nails with neoprene washers under the head are made for plastic panel installation and are recommended by panel manufacturers. Panels should be nailed every 12 inches. It is important that the nails be driven through the crowns, not through the valleys of the corrugations. Don't apply the hammer with too much zest or drive the nails in too far, for if you do, you may deface the crown.

Special wood screws are available that may be used instead of nails. Whether you use nails or screws, be sure to apply mastic sealant first (see drawing at left).

If you are attaching your panels to metal framework, use self-tapping screws or bolts inserted through holes that have been drilled through the crown and into the metal. The bolt should be drawn up snugly but not too tightly.

Non-setting mastics are recommended for sealing the laps between panels; they should be used in all outdoor installations to make the joints watertight. (Some users have found that hard-setting sealers will eventually crack under minute flexing of the panels caused by wind.)

Beware of the heat trap

Any solid, translucent roof attached to two or more walls of a house without ventilating provisions, is a potential heat trap. Along with light, heat radiates through, and if the air doesn't have a chance to circulate, there may be an uncomfortable heat build-up.

Special flashing, end strips, and other accessories make installation easier and more permanent. The need for these extras varies with the type of installation and sometimes with your climate.

The illustrations below show how the panels may be installed to permit free air circulation. More than one patio cover has been rebuilt because it cooked rather than cooled those who sat beneath it, so it would be advisable to study the methods suggested for providing ventilation.

TWO WAYS TO VENTILATE

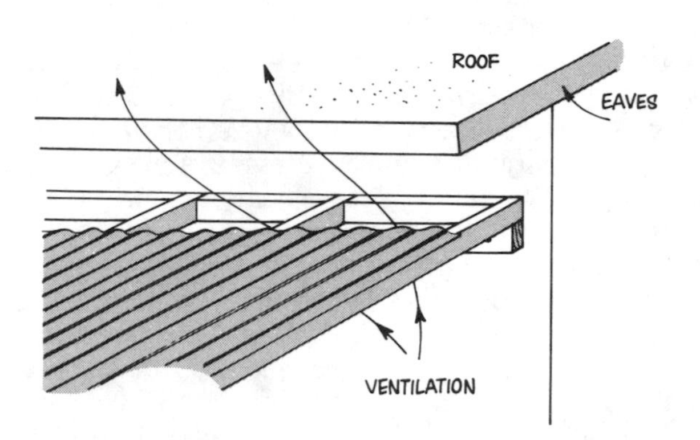

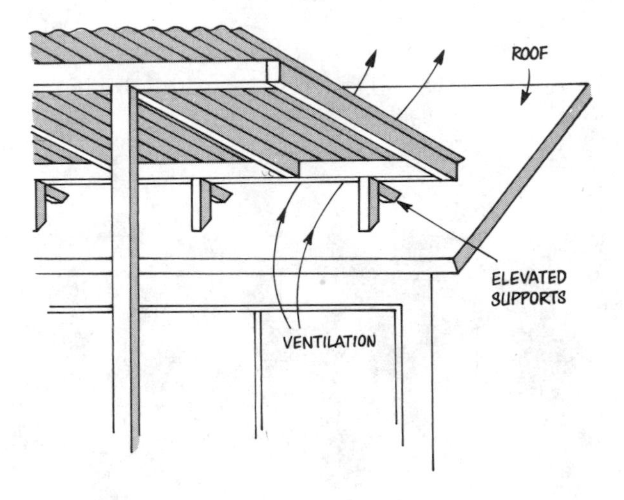

Since there is a wide range in thermal conductance among the plastics, choose one with a low percentage of heat transfer, if relief from the heat is what you're after. The amount of light that filters through the panel is not necessarily a strict index to the amount of heat that will also be transferred, but in the absence of other data, light transmission is as good a clue as any.

If you're building the overhead above plantings, you should take extra care to provide adequate ventilation. Without it, condensation of moisture on the under surface of the plastic may occur, causing an annoying drip problem.

RENEWING A PLASTIC ROOF

When those panels of fiberglass-reinforced plastic on your patio roof or garden fence get dirty, rough, and dull looking, you can improve their appearance dramatically by cleaning them and renewing their smooth finish. In the 25 years or so that these panels have been on the market, we have discovered the need for such maintenance.

In unsheltered locations, the surfaces of these panels will roughen six months to several years after you install them, depending on panel quality and weather (temperature extremes, sunlight, rain, sand-laden winds, and chemicals in the atmosphere). The plastic binder deteriorates at the surface and exposes glass fibers; then the roughened surface traps dirt, and the panels darken. Some readers report that birds further damage worn panels by pulling out loose fibers to make their nests.

You can help put off the time when you'll need to resurface if you will just hose the panels clean once every three weeks or so. And you'll have less work to do if you resurface the panels as soon as they become dull-looking.

To prevent damage to the roof, do your walking on lengths of inch-thick board set across the corrugated paneling. (Be sure the roof structure is strong enough to support your weight; it may need reinforcing.)

SAME SECTION of fiberglass-reinforced plastic roof before and after cleaning with water and steel wool.

Start cleaning at the high end of the panels and work down in the direction of their drainage, flooding with water from a hose and cleaning with large pads of fine-gauge steel wool, working lengthwise along the corrugations. (If your hose won't reach, you can use the steel wool without the water.) The steel wool cuts off exposed glass fibers and scrapes out imbedded dirt.

Avoid using soap or detergent on panels, since even a trace of it can adversely affect the bond of the resurfacing lacquer.

Work carefully—any dirt you leave will be trapped by the finish and will show when you look through the panels from the other side.

Be sure you let the cleaned panels dry completely, since moisture (like soap) can interfere

EASY WAY to clean panels. Flush surface with water only (no detergents); work the fibers and dirt loose.

SPRAYED SURFACE FIBERS disappear under application of lacquer. Drying takes half an hour.

with the bond. Replace old nails and fasteners (if necessary) before you begin any resurfacing.

Applying the plastic finish

Finish is a formulated liquid resin that dries to a tough, smooth, protective skin often stronger than the original panel surface. Known by such names as surfacing lacquer and resurfacer, it is usually applied in clear form over panels of any color (more than 99 percent of the refinisher sold is clear). You can also get it with color added.

If you're applying clear lacquer to a small area, a brush works well. Fresh lacquer dissolves a recently applied coat, so work toward the already covered portion, lifting your brush as it gets to this part.

If you are clear-finishing 50 square feet or more of paneling or applying colored lacquer, you'd better use a sprayer. You will get a more even application without telltale brush streaks. Hold the nozzle 12 to 18 inches from the panel surface and apply a liberal coat, working with the corrugations. (Caution: if you overcoat the panels, the lacquer will bead and run.) Sprayers can be rented inexpensively at tool rental shops and paint stores; surfacing lacquer is available in quarts and gallons where panels are sold. A quart covers about 110 square feet. You should be able to clean and refinish at least 150 square feet of paneling in an afternoon.

When spraying, thin the lacquer with ordinary lacquer thinner or with toluene (available at chemical supply houses); mix one part thinner to four parts lacquer. Use thinner to clean up sprayer or brushes.

One or two coats of lacquer give panels a resistant surface that prepares good quality panels for another 5 to 6 years of average weathering. Future coats of finish, when needed, should maintain these panels indefinitely. The lacquer can resurface but cannot restore "bargain" panels made from inferior plastic mixes with dyes rather than from permanent pigments; these tend to fade or change color as they age.

To cut down slightly on the heat and light coming through too-translucent panels, use a colored lacquer. Spray evenly to avoid making it appear blotchy from beneath.

White is more effective than any other lacquer color because it reflects more sun and keeps the under-roof area cooler.

Before you apply any of the several colored lacquers, paint some on an extra piece of panel to test its effect on light transmission and color. If you want to leave the conditions of light and heat transmission unchanged, use clear lacquer.

CORRUGATED ALUMINUM PANELS

Aluminum is the ideal material for outdoor construction. It can't rust, rot, or warp, is easy to handle and economical. Precision manufacturing provides narrow width in standard panel lengths. Aluminum is so lightweight that installation requires very little muscle power.

A complete set of installation instructions is usually packed with the panels by the manufacturer. Also provided is a complete list of accessories—ridge cap, end wall flashing, rubber filler strip, nails, and caulking compound. By planning your design carefully and discussing the project with a knowledgeable and trustworthy building supply man, you can avoid mid-project problems. Also discuss your project with the building inspector. Each locality has a few specific guidelines that will help you to build a long lasting structure.

Installation of corrugated or crimped aluminum panels is accomplished in the same manner as corrugated plastic panels. *(See How to install corrugated panels, page 58.)*

Panels are available smooth or in embossed patterns. Besides the conventional corrugation, there are V-crimp and "flat-top" rib configurations. You may choose from several colors: white, gray, blue, green, gold, and red. If color is no special factor in your plans, the natural aluminum (less expensive) is also very popular.

The panels come in widths that will give a 24-inch coverage with lap. Color panels come in lengths of 8, 10, 12, 14 and 16 feet; plain panels vary from 6 to 24 feet. They may be purchased from building supply stores, some hardware stores and lumber yards, and mail order houses.

Aluminum is easy to cut across the corrugations with tin snips or a power saw (use an abrasive blade) if you should need a special length. Some products can also be cut lengthwise by merely scoring wtih a sharp knife along a straight edge and breaking along this line.

Use aluminum nails with neoprene washers when attaching panels to the overhead frame, figuring 100 nails for every 100 square feet (colored nails to match panel colors are available from your dealer). Be sure to nail through the crowns of the corrugations, not through the valleys.

Aluminum panels are excellent for reflecting heat. They bounce it back into the sky in the daytime. At night or on cool days, if you have a patio warmer, they will reflect heat back into the patio, adding hours of comfortable outdoor living.

Many people like to combine the opaque aluminum panels with translucent plastic panels for a roof that lets in some light but little heat. This can result in a handsome and effective overhead.

EXTRAORDINARY living room features dining, swimming, bar, garden work center, hanging gardens. Sliding doors at end open to terrace and hillside garden. Electrically controlled roof (see page 63) is in closed position.

LEFT. Plastic panels in movable roof mounted on corrugated wood strips and set in western cedar frame. ABOVE: stained redwood and plastic covered eggcrate work well with shoji in Japanese garden and pond.

The ventilated patio

SWITCH shown in operation at right opens roof over pool to any desired position. When roof is completely open, far end of pool receives more shade from doubled overhead sections. (Architect: Dean Price.)

ABOVE. Large patio has overhead 16 feet square. Flat sheets of fiberglass cut down glare from the sun.
LEFT: gravity operation opens glass paneled overhead. Room is lanai with full length interior garden.

Louvers for sun or shade

Preassembled metal units and do-it-yourself wood projects

SUN SHELTER of 1 by 6s in three 4 by 8 adjustable panels.

Do you want a permanent, airy, tailored patio shelter that will block the last shred of direct sunlight at noon or in the morning or afternon, yet let the light pour through at other times of the day? Maybe a handsome louvered overhead will fill the bill.

Adjustable louvers can give you almost any degree of light or shade you may want through the day, whereas fixed louvers can be designed to block the sun during that part of the day when the sun is unwanted.

A louver shelter is like a lath cover in that it is made up of parallel boards, but the boards are set on edge or at an angle to take the greatest advantage of their width in blocking the sun.

Obviously, more thought must go into planning a fixed louver overhead than is required for designing an adjustable system, for once the boards are nailed in place, you'll have to live with them as they were designed.

Like all overheads, the louvered species cannot do anything about the light and heat that will fill the area beneath when the sun is low in the sky. All overheads need help from some vertical screen —fence, trees, or drapes hung from the overhead frame—to keep the eastern morning sun or western afternoon sun from thwarting your efforts to build a glare-free, cool retreat.

PLANNING AND BUILDING LOUVERS

Generally speaking, if you run your louvers east and west, slanting the boards away from the sun, you will block the midday sun and may admit morning and afternoon sun. If you run them north and south, you will admit either morning or afternoon sun, depending on the louvers' slant.

HANDCRAFTED LOUVERS can be adjusted by rope and pulley at each post.

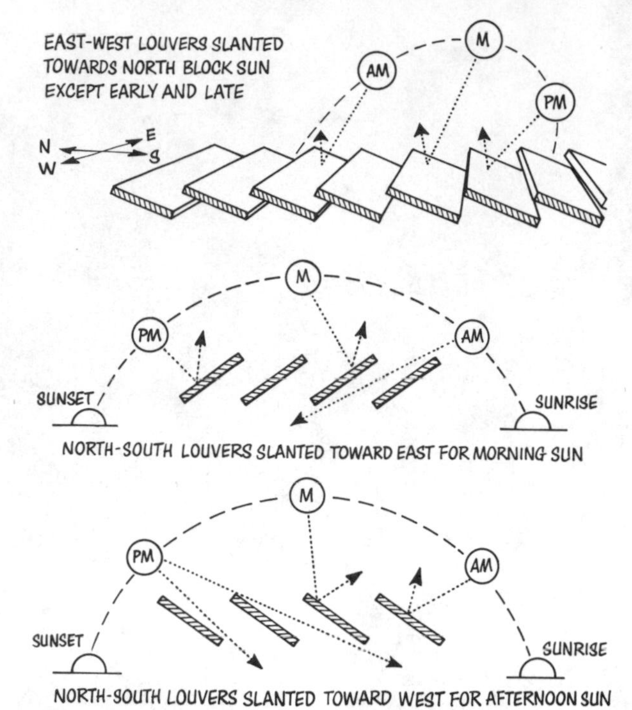

In most cases the house against which the overhead is to be attached will not run exactly with the cardinal points of the compass, but with a little thought, regardless of the orientation, you should be able to build a louvered overhead that will effectively protect the patio area when you want it protected.

Louvered overheads can be used very effectively to control indoor temperature and often are welcome solutions to the very tough western exposure problem.

If you plan adjustable louvers that can be completely shut, the sun problem is more easily solved; but even adjustable louvers will function better if you think through the specific problem the roof is supposed to solve.

With adjustable louvers you can block the noonday sun by having the boards run north and south and being completely closed, but if the boards run east and west, they can be half opened to let reflected light through while still blocking all direct rays.

Setting the angle

With adjustable louvers, of course, you don't have to worry about setting the angle of the boards while you're planning; but if you are building a shelter with fixed louvers, it's well worth analyzing how the angle will affect the light and shade you will get.

First, remember that in summer, the sun is higher in the sky than through the rest of the year. Since the louvered overhead will be designed to keep direct light out part of the day, you have to consider what the height of the sun will be during summer at that time.

For example, if you wish to block the sun at noon on a southerly exposure with louvers running east-west, you'll have to build them to keep out a sun that is almost overhead in Southern California and Georgia, but you'll have less to contend with in Washington and Maine.

The sun's altitude at 8 a.m. and 4 p.m. is within 1° of 37° in all the latitudes in the chart.

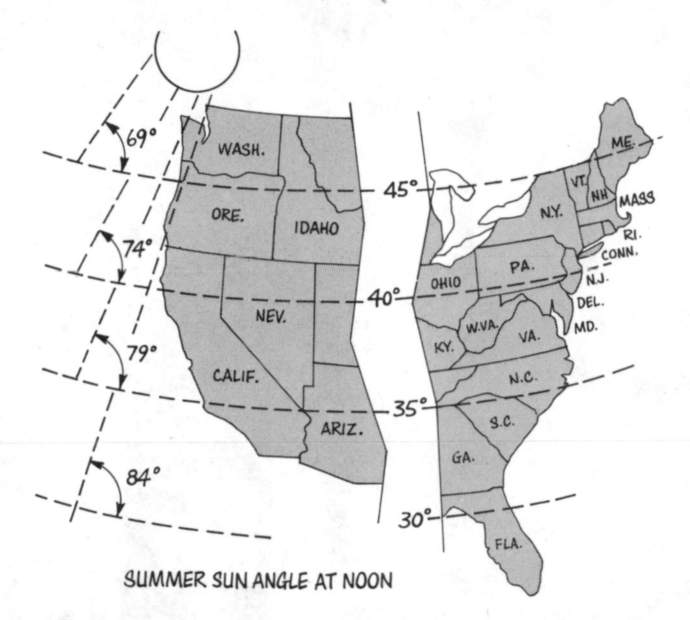

SUMMER SUN ANGLE AT NOON

Let's assume that you live at a point where the maximum sun angle is 75°. There are several ways to set the angle, and, as you can see, there is quite a difference in the amount of lumber required and in the amount of reflected light that can shine through.

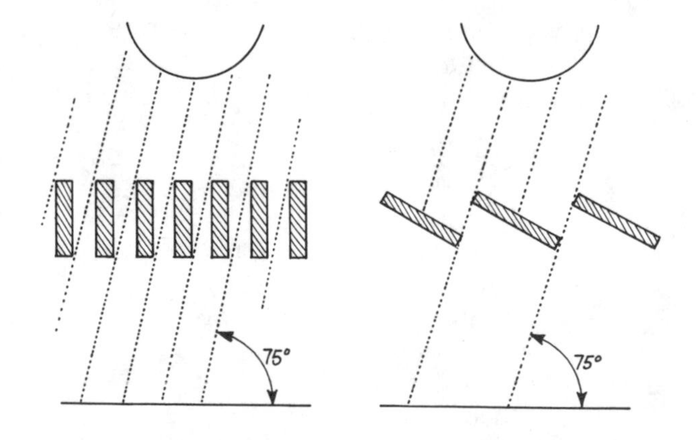

The more you tilt the boards, the fewer pieces you will need, but if you try to spread them too far, you will diminish the amount of reflected light that can shine through.

If you plan to have the plane of the overhead at any angle other than horizontal, add the angle of the pitch to the angle of the sun's altitude that you're trying to beat.

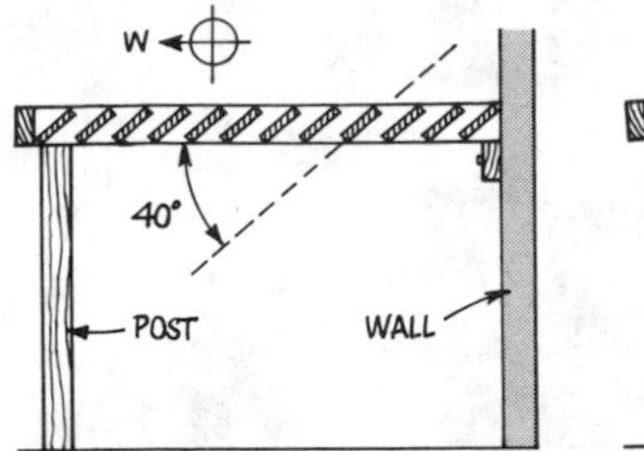

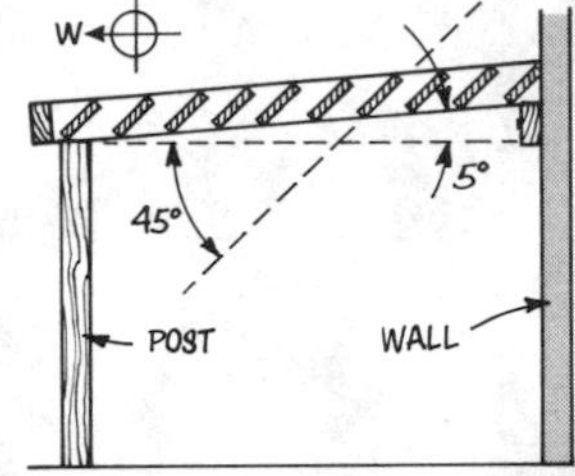

Building a fixed louver

Once you have settled on the orientation, angle, and spacing of the louvers, you have a choice of building the cover of removable panels or attaching the louver blades right to the framework of the overhead. In either case, one of the easiest and surest ways of doing a precision job is to cut spacers of the exact dimensions to fit between each louver.

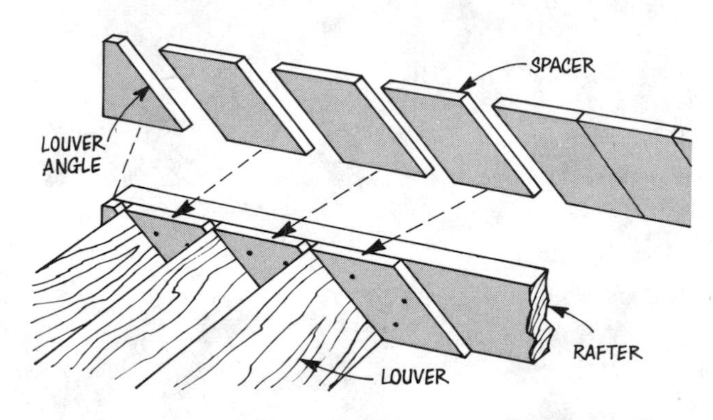

Nail the first louver in place, using the triangular piece as the first guide, and then nail the rest of the louvers and the sets of parallelogram-shaped spacers in turn to the frame.

Other ways to do it: 1) cut the rafters in steps and then nail the louvers directly to them. Be sure the dimensions of the rafters measured through the depth of the cut are those specified for the span.

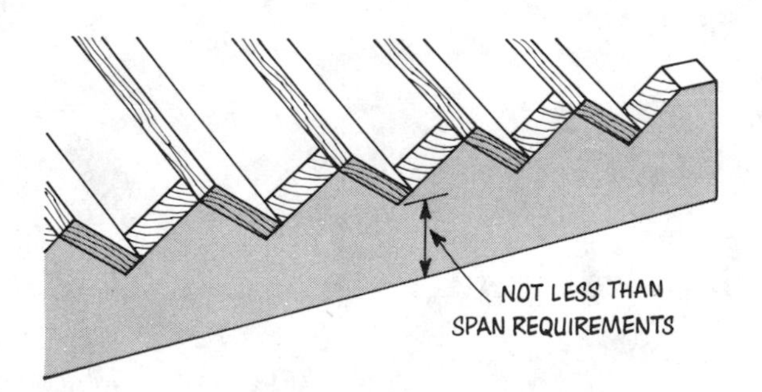

Or 2), cut steps in separate boards that can be attached on each side of the rafters to support the louvers.

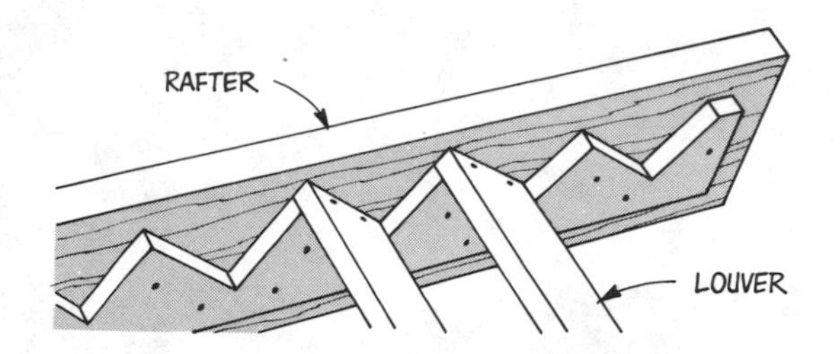

To avoid warping you should use boards 1 inch thick and the span they cover should not exceed 4 feet.

The width of the boards can vary, but the narrower they are, the closer they'll have to be spaced. Four or 6-inch boards are practical.

If you want more reflected light, paint one or both of the louvered surfaces white.

Building an adjustable louver

No overhead you could build will take more patience and good workmanship than an adjustable louver, but except for the fact that it's not waterproof, there's probably not a more generally satisfactory cover that can be made. You can buy them ready-made in a variety of materials if you do not want to attempt the precision workmanship required by such a structure.

The plans which follow show you how one builder did it. The simplicity of this design and the efficient way it works to give you shade when you want it make it a cover hard to beat.

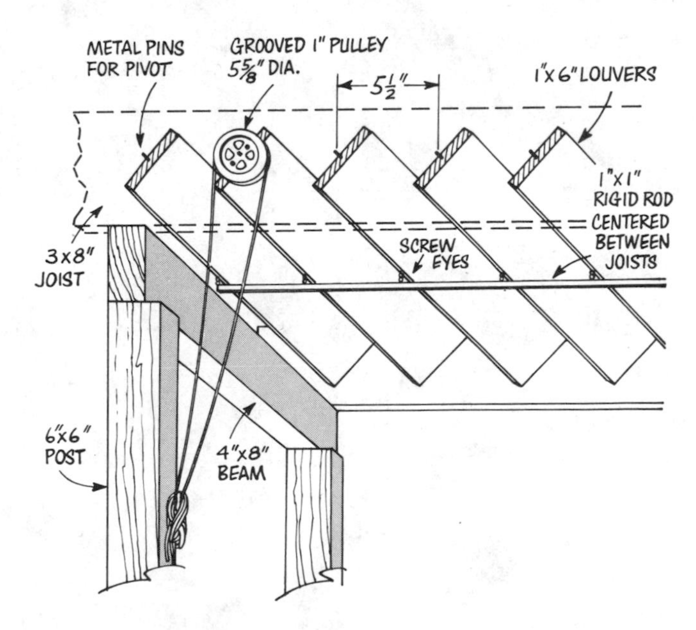

OPEN AND CLOSED louvers here show louver sun and shade control.

LOUVER LEVER with rod attached to each louver allows single motion opening and closing. Hardware, building supply dealers sell parts.

Controlling sun and shade

LOUVERED ROOF and graceful latticework arches turned this once unprotected back deck into an airy gazebo. Design provides privacy and relief from the afternoon sun.

MOUNTED on manufactured metal rafters and metal strip beams, aluminum louvers are the stationary type.

NOTCHED into 4 by 6 sculptured beams, 1 by 8 stationary louvers keep sun from the living room.

PARALLEL BEAMS simultaneously support upper row of beams while serving as cross-brace for support posts.

Eggcrate, the versatile roof

For climbing vines and hanging gardens

ROOF POSTS were spaced for uninterrupted view of garden.

Some patios need all the open sun they can get, yet beg for the sheltered feeling that an overhead roof can provide. If you are faced with this dilemma—sun versus shelter—there is a compromise that may solve your problem. An eggcrate shelter, that is open to the sky but substantial enough to give the feeling of protection, may be your answer.

If your patio is on the north side of the house where the sun seldom enters or on the eastern side where only the morning sun can enter freely, or if trees surround the patio and block the sun, an eggcrate overhead will let in most of the valuable sunlight, cutting only the slanting rays of early morning and late afternoon sun. It will also deflect the wind a bit.

SHADOWS, VINES AND SUNSHINE

The shadow pattern of an eggcrate shelter is strong but will not afford an appreciable amount of relief from light or heat when the sun is high. If you are trying to escape light or heat, you had better build a denser overhead, cover the eggcrate with one of the lighter materials or start a deciduous vine up the posts to eventually give the summertime shade you want.

Eggcrate overheads, by their very design, are natural showcases for climbing vines. And for many gardeners, a patio that is roofed at least partially with vines has certain advantages that you never get if you cover it with structural materials only.

Most of these advantages derive from the nature of plants themselves. Plants are super air conditioners. Their leaves transpire moisture on hot days, cooling the immediate surrounding atmosphere. Unlike many structural materials, they do not re-radiate heat.

There's aesthetic delight in the rustling sound of wind passing through foliage. Leaf colors are refreshing to the eye. Many vines offer a fascinating study in branch and leaf pattern, in flower form and color.

EGGCRATE ROOF DESIGN is handy for climbing vines, hanging baskets.

VINES FOR PATIO ROOF COVERINGS

Here are some of the vines which have been used successfully on overheads.

The figures in parentheses following the vine names are the minimum temperatures the plants can withstand.

Anemone Clematis *(C. montana)* (0°). Deciduous. Moderate to fast growth. Medium to dense shade. Mass of white to pink blossoms in early spring. Vigorous climber with light green, divided leaves. Prune lightly. Variety *m. rubens* often preferred for pink to rose flowers.

Blood Red Trumpet Vine *(Phaedranthus buccinatorius,* sometimes sold as *Bignonia cherere)* (28°). Evergreen. Rapid growth. Medium shade. Vine will bloom brilliantly whenever weather is warm.

Boston Ivy *(Parthenocissus tricuspidata)* (0°). Deciduous. Rapid growth. Dense cover. Glossy, bright green, ivy-like leaves. Excellent fall color in all but very mild winter areas.

Bougainvillea (20°). Evergreen. Moderate to rapid growth. Medium to dense shade. Brilliant purple, red, magenta flowers in summer. Severe frost may kill. Medium green foliage.

Burmese Honeysuckle *(Lonicera hildebrandiana)* (25°). Evergreen. Fast growth. Moderate shade. Dark green, glossy leaves. Fragrant, creamy white blossoms in early summer; blossoms turn yellow as they age. Needs lots of water.

Carolina Jessamine *(Gelsemium sempervirens)* (25°). Evergreen. Moderate growth. Light shade. Yellow-green, glossy foliage. Small yellow flowers in early spring. Trains well; foliage cascades downward. All parts of plant are poisonous, including flower nectar.

Cats Claw or **Yellow Trumpet Vine** *(Doxantha unguis-cati)* (30°). Evergreen, deciduous in cold winters. Rapid growth. Medium shade. Glossy green foliage. Likes hot sun. Bears yellow flowers in early spring. Prune severely after blooming.

Clematis jackmanii (0°). Deciduous. Rapid growth. Very light shade. Soft green divided leaves. Large purple flowers freely produced in summer. Cut back dormant stems after new growth starts in March.

Common Jasmine, Poet's Jasmine *(Jasminum Officinale)* (15°). Evergreen, semi-deciduous in cold winters. Rapid growth. Light shade. Fragrant white flowers bloom through spring. Needs pruning, training through growing season. *J. o. grandiflorum* popular.

Evergreen Clematis *(Clematis armandii)* (15°). Slow to start, rapid growth later. Light to medium shade. Dark green, glossy foliage. White, star-like flowers in spring. Vigorous growth needs attention to keep in check. Won't bloom some years.

Grapes (American, 0°; European, 20°). Deciduous. Rapid growth. Dense shade. Luxuriant foliage makes cool shade. Fruiting varieties provide eating grapes but also bring flies and bees. Select grape that will not fruit in your locality if you want shade alone. Yearly pruning almost necessary.

Ivy *(Hedera).* Evergreen. Slow to start, rapid growth when established. Dense shade. English Ivy *(H. helix)* (0°) and Algerian Ivy *(H. canariensis)* (20°) both used; Algerian Ivy takes hot sun better. Dense, glossy foliage. Use variegated forms for brighter foliage. Should be cut back annually.

Japanese Honeysuckle *(Lonicera japonica)* (10°). Evergreen. Very rapid growth. Moderate to dense shade. Dark green foliage. White, purple tinged flowers bloom in late spring. Hall's Honeysuckle (L. j. *halliana)* usually planted. Needs constant pruning. Don't plant near sewer and drain lines.

Passion Vine *(Passiflora).* Semi-evergreen. Rapid growth. Dense shade. Rank grower, needs ruthless pruning, much water. Passion fruit *(P. edulis)* (28°) will yield edible fruits in mild sections. Hardiest species is *P. alato-caerulea* (25°).

Roses (0°). Deciduous. Usually rapid growth. Light to medium shade. Many varieties. Choose disease-resistant, long-blooming type. Select variety with good quality foliage. Prune severely for best results.

Sweet Autumn Clematis *(Clematis dioscoreifolia robusta,* formerly *C. paniculata)* (0°). Deciduous. Fast growth. Medium to heavy shade. Blooms profusely in autumn. Fragrant white flowers borne on new growth.

Violet Trumpet Vine *(Clytostoma callistegioides,* often sold as *Bignonia violacea)* (20°). Evergreen. Rapid growth. Medium to dense shade. Violet-lavender to pale purple flowers in late spring to fall. Sprays of flowers and foliage hang downward. Slow growth first year.

Virginia Creeper *(Parthenocissus quinquefolia)* (0°). Deciduous. Rapid growth. Medium shade. Sends out light, drooping branches which wave in breeze. Medium green leaves divided fan-wise into leaflets. Excellent autumn color in all but mild winter areas.

Wisteria (0°). Deciduous. Fast growth. Medium to heavy shade. Beautiful vine. Fragrant, pendulous clusters of white, lavender or purple flowers in spring. Beauty increases with age. Foliage light green in summer. Attractive branch pattern in winter.

If you live in a two-story house or on a sloping site, you enjoy the exciting advantage of being able to look down on a vine-covered roof, as well as being able to look up into it from below. On such sites, you can reap full pleasure if you plant a vine that carries its flowers on the top.

Some gardeners will always want to plant vines because they like to watch and listen to birds, bees, butterflies, and other creatures attracted by lush foliage, by the color and nectar of flowers, and—on some vines—by fruit.

Of course, vines also have disadvantages. Many vines are slow growing (you can offset this disadvantage by providing a temporary covering such as burlap or lath or by planting a quick annual vine along with a permanent one). Upkeep is a bugaboo—you'll have to train, prune, and spray. All vines, evergreens included, drop leaves to some degree at one time or another. The fact that vines lure birds, bees, and wasps must also be taken into consideration.

DESIGN AND CONSTRUCTION

The construction of an eggcrate shelter is simple. Once the basic structure is built, all you have to do is nail in blocking between the rafters. There is a bit of skill required to nail the blocking exactly in place, but once the technique is mastered, it's just a matter of sawing the blocking to size and nailing it in place.

Materials

Any commonly available, structural soft wood is suitable, but unless you use the heartwood of redwood or western red cedar, you will have to protect it from the weather with paint, preservative, exterior grade varnish, or other sealer.

Use 10 or 12-penny galvanized or aluminum alloy nails with 2-inch lumber, 8-penny with 1-inch boards.

Structural requirements

A common temptation is to build an eggcrate that is just strong enough to support itself. When the builder later decides to cover it over, he may be disappointed to learn that his structure must be partly rebuilt to stand up under the added weight and wind load created by closing over the surface. You would be wise to build it to the specifications for roof framing and support outlined in the materials chart, page 19. The blocking that you add to the basic structure in the chart will not affect the roof, and you can later close it over if you desire.

In general (following the chart), if your blocking is the same size lumber as the rafters to which you nail it, and you space the blocking approximately as far apart as the rafters, you can be sure the structure will support its own weight and more without sagging.

For example, if your rafters are 2 by 8s spaced 24 inches on centers, you can nail 2 by 8-inch blocking on 24-inch centers without fear of overloading the structure.

The blocking need not be of the same dimensions as the rafters, but for structural reasons and for appearance, never use blocking of greater dimensions than the rafters.

An eggcrate cover with rafters and blocking on 1-foot centers can be made of 1 by 6-inch boards, but use 2-inch lumber for the rafters if you space them wider than 1 foot apart.

How to install blocking

One easy way to make sure the blocking will line up and appear to run through the rafters as a single piece is to measure off and mark the spacing you want on the two end rafters. Then stretch a chalk line between the marks and snap it to mark the rest of the rafters. Repeat this for each row of blocking. Use a square to draw center lines down the vertical faces of the rafters so that you can see them when you are working from below.

When you're ready to nail the blocking in, carefully measure the distance between the rafters and saw the blocking to fit as closely as possible. Cut the blocking to fit snugly, if possible, but don't bow the rafters by using pieces too long.

Use four nails to secure each end of the blocking, two on each side. Offset the nails a bit to avoid splitting the lumber.

Before driving any one nail all the way in, drive one nail from each side of the blocking into the rafter so that you can tap the blocking into the exact position you want before making it permanent.

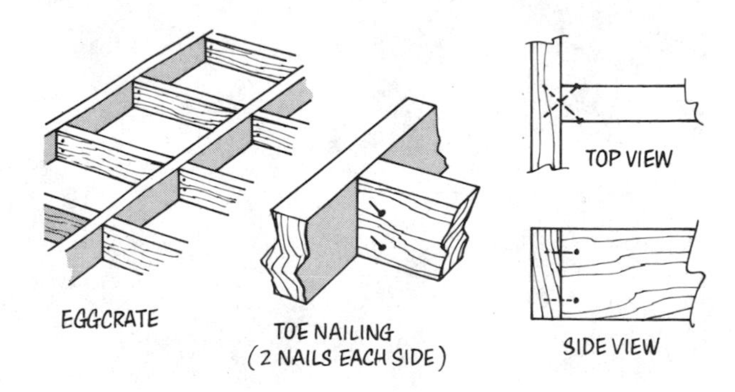

The last few blows you give the nails as you toenail them in are liable to mar the wood. Unless you are an expert, it might be wise to use finishing nails and drive them with a nail set before the hammer marks the blocking.

Versatile eggcrate

COVERED EGGCRATE sections give partial shade to semi-tropical garden and hanging plants.

LINED EGGCRATE sections of 1 by 6 lumber and finished lath and 1 by 2 batten give shade, sun control.

LEFT: extensive use of screened and unscreened eggcrate design. White paint reflects hot sun. Planting area was converted to experimental garden. RIGHT: dramatic uses of eggcrate construction.

SPACED REDWOOD batten panels allow ventilation, can be moved about in openings to control sun and shade, depending upon the season.

REDWOOD is dominating feature; is used in eggcrate overhead, garden furniture, deck.

A solid roof adds permanence

Indoor-outdoor rooms and tea houses

NATURAL, unfinished redwood with streaks of sapwood.

We are concerned here with overhead construction materials—both permanent and temporary—that will thwart the sun altogether and keep you dry when it rains. The materials in this chapter, however, are not the only ones that can be described as "solid." Plastic, for example, makes an excellent solid overhead—and yet, because of its unique ability to transmit light, we have covered it in a separate chapter.

A DESIGN FOR PERMANENCE

It is important to check plans with your local building department and obtain a building permit before undertaking any solid roof covering. Roofing your patio may be the first step in enclosing a room as part of your house. Because of this, its design should be taken seriously. The generally recognized structural specifications given in the front of this book should be followed if you want the overhead as a permanent part of your home, but there may be local variations in allowable construction and pitch of the roof (all solid roofs require some slope to take care of the water runoff).

Materials

As suggested on page 10, solid patio roofs have a way of becoming completely enclosed rooms. When planning your patio roof therefore, it is wise to look into the building code requirements for adding a room to your house. A little extra time spent now can save you considerably more time (and money) later.

Galvanized iron. Heavier and harder to work with than aluminum, galvanized iron nevetheless makes a rugged, economical patio cover. If you paint it,

GRECIAN-LIKE COLUMNS of redwood lumber enhance semi-formal and rustic atmosphere of terrace.

it will last indefinitely. The corrugated sheets are 27½ inches wide and come in lengths of 6, 8, 10, 12, and up to 20 feet. Be sure to use galvanized nails with this material.

Asbestos cement board. This material is practically indestructible, and, with a good coat of paint, makes an attractive cover that is "here to stay" in appearance as well as in fact. It is available in flat sheets or corrugated; if you use the latter, be sure to nail through the crowns at the corrugations, rather than the valleys. Overlap the sheets 1½ corrugations and seal with mastic.

HOUSE ROOF and ceiling protect poolside lanai. North wind can be blocked by sliding wooden wall section.

SLIDING PANEL drawn from niche closes off lanai. Bolt on post keeps panel from blowing sideways.

Roofing paper and shingles. The most common way to make a permanent roof is to cover the rafters with 1 by 6 or 1 by 8-inch sheathing and put down composition paper. Along with each roll, you'll receive mastic to seal the seams and galvanized roofing-nails for the edges.

The paper roof should last five to ten years, depending on the climate. As an optional step, tar and gravel may be brushed on the tar paper.

Shingles and shakes are easy to install. You start at the bottom of your roof's slope and work across, nailing each shingle with two or three galvanized nails. Nail down the next row, making sure to cover each open space between shingles in the first row. Leave 5 inches of shake or 3½ inches of shingle in the first row exposed. Work right on up following the same procedure.

TEMPORARY SOLID ROOFS

Practically any solid, flat, opaque material that you can name is a candidate for part time use as an overhead. Exterior plywood and tempered hardboard are good materials to use as panels over your shelter. Check with your lumber dealer for sizes that are available. Ask him what he charges for specially cut material if he doesn't have what you want. Standard size for plywood is 4 by 8 feet, but lumber yards that cater to the weekend trade may be able to give you any sizes you want. Figure the modules as even parts of the 4 by 8-foot sheet.

Tempered hardboard comes in 4 by 8 sheets, but some outlets have available ready-cut sections in 4 by 1, 4 by 1½, 4 by 2, 4 by 3, 4 by 4, and 4 by 6-foot sizes.

Perforated hardboard, which lets slender pencils of light through, is usually available in the following sizes: 2 by 3, 2 by 4, 2 by 6, 2 by 8, 4 by 3, 4 by 4, 4 by 6, and 4 by 8-foot sheets.

Plywood (be sure to specify *exterior* grade) and hardboard must be painted, varnished, sealed, or otherwise treated to keep the weather from the material. It is important to paint the under side as well as the exposed surface. Be sure to seal the edges of plywood with at least two coats of sealer.

Temporary solid roofs should be built so that panels can be stored indoors when not needed for shade.

DISAPPEARING DOORS create open feeling in this heated Japanese bath. Interior hanging garden receives sunlight from windows at upper right and open roof above. Exterior hanging garden is at left. (Design: E. H. Nunn.)

A convertible lanai

FITTED SLIDING PANELS are lowered on pulley attached inside lanai. Sections slide on track.

Interior view.

INDEX

A-frame, 11
Aluminum panels, 61
Asbestos cement board, 78

Bamboo
- overhead, 27
- types of, 28

Basket-weave covering, 50
Beam size, how to figure, 18
Beams, installing, 22-23
Blocking, 21
- installation of, 73

Box beams, making, 28-29
Bracing, how to do, 23
Building
- code, 15
- permit, 15

Canvas
- maintenance of, 48
- mildew, 48
- protection of, 48
- types of, 47-48

Climate, 6
Construction costs, 10
Corrugated panels, 58
Covers
- adjustable, making, 30
- bamboo, 27
- canvas
  - adjustable, 49
  - basket-weave, 50
  - framework for, 50
  - lace on, 49
  - making (sewing), 49
- construction, 24
- patio
  - asbestos cement board for, 78
  - galvanized iron for, 77
- reed, 27-28
- synthetic fabrics, 48
- vines for roof, 72, 73

Cross-bracing, 21

Eave, connection to, 19-20

Eggcrate
- construction of, 73
- designs with, 73
- material for, 73
- placement of, 71

Fabrics, synthetic, 48
Facing strip, 20
Flyscreen, 33
Footing requirements, 23
Footings, types of, 17
Framework, pipe, 50

Galvanized iron cover, 77
Glass
- panels, 53-54
- types of, 55

Insect protection, 34

Lath
- building frame for, 43
- buying, 43
- construction, 40-42
- designing with, 39
- requirements, 42-43
- spacing, 43
- types of, 41
  - batten, 41
  - grapestakes, 41
  - lathhouse, 41
  - utility fencing, 41

Ledger, 20
- position of, 21

Louvers
- adjustable, 67
- fixed, 67
- location of, 66-67
- planning for, 65
- setting the angle, 67

Lumber, length of, 18

Materials
- requirements, 19
- summary of, 13

Overhang, estimate of, 21

Panels
- aluminum corrugated, 61
- buying plastic, 57
- cleaning, 60
- corrugated, 58
- glass, 53-54
- installation, 58-59
- plastic, 57
- plastic finish, 61

Panes, setting, 55
Patio
- location of, 8
- L-shaped, 10
- U-shaped, 10

Permits, building, 15
Planning, 5
Plants, roof covering, 72-73
Plastic
- colors of, 57
- sizes of, 57
- types of, 57

Posts
- anchors, 16
- and beams support, 17
- erection of, 23
- pipe installation, 25
- support, 16

Radiant heat, 56
Rafters
- installing, 21-22
- pitch of, 21
- sizes, how to figure, 18
- types of, 17

Rain, effect of, 8
Reed
- installation, 28
- overhead, 27-28
- rolls, types of, 27-28

Roof
- cold, 11
- connection to, 20
- construction, greenhouse type, 54
- designs, 11
- flat, 11
- hip, 11

(Cont'd.)
- no-eave, 11
- plastic, renewing, 60

Roofing materials, 78
- chart, 12
- types of, 78

Roofs
- future planning, 77
- paper and shingle, 78
- temporary solid, 78

Screen
- materials, 34-36
  - aluminum-&-plastic, 35
  - aluminum, 34
  - fencing, 36
  - galvanized steel, 35
  - glass fiber, 35
  - louvered, 36
  - plastic, 35
  - saran, 36, 48
  - wire mesh, 36
- roof, cleaning, 34

Screens
- temporary, 7
- vertical, 7

Skylight, construction of, 54
Snow
- effect of, 8
- loads, 11

Steel frames, 23
Sun
- controlling exposure, 69
- exposure, 6-7
- protection, 42
- louver, 66

Synthetic fabrics, 48

Ventilation, 10, 56
- patio, 63
- ways of, 59

Walls, connection to, 20
Weather
- conditions, 6
- information, 6

Wind, effect of, 8

Photographers

Cover photograph by **George Selland. Jerry A. Anson:** page 69 (bottom right). **William Aplin:** page 46. **Nancy Bannick:** page 31 (top left). **Ernest Braun:** pages 30, 51 (bottom), 63 (bottom right), 64, 69 (bottom left), 75 (bottom). **California Redwood Association:** pages 40 (left), 62 (bottom right), 74 (bottom). **Glenn Christiansen:** pages 9 (bottom right), 31 (bottom right), 37 (bottom left), 52, 55, 63 (bottom left), 68 (bottom), 74 (top right). **Richard Fish:** pages 4, 32, 33, 36, 37 (top), 38, 44 (top), 45 (top left, bottom), 75 (top right), 77. **A. L. Francis:** page 69 (top). **Bert Goldrath:** page 29. **Hoover Canvas Products Company:** page 51 (top right). **Ells Marugg:** pages 9 (top left) 14, 15, 31 (top right, bottom left), 39, 44 (bottom), 62 (top), 63 (top left, right), 74 (top left), 75 (top left), 79. **Michael McGinnis:** page 60. **Ken Molino:** page 9 (bottom left). **James F. Morgan, Sr.:** page 26. **Phil Palmer:** page 51 (top left). **Photo-Craft Company-Hawaii:** page 78. **Norman A. Plate:** pages 53, 62 (bottom left), 76. **Martha Rosman:** page 37 (bottom right). **Blair Stapp:** page 68 (top right). **Van Nuys Awning Company, Inc.:** page 47. **Darrow M. Watt:** pages 7, 9 (top right), 27, 45 (top right), 56, 65, 68 (top left), 70, 71. **Western Wood Products Association:** page 40 (right).